Crosswalk Coach for the Common Core State Standards

Mathematics

Grade 6

Crosswalk Coach for the Common Core State Standards, Mathematics, Grade 6
301NA
ISBN-13: 978-0-7836-7850-4

Contributing Writer: Randy Green
Cover Image: © Ron Hilton/Dreamstime.com

Triumph Learning® 136 Madison Avenue, 7th Floor, New York, NY 10016

Printed in the United States of America.

10 9 8 7 6 5 4 3 2

Frequently Asked Questions about the Common Core Standards

What are the Common Core State Standards?

The Common Core State Standards for mathematics and English language arts, grades K–12, are a set of shared goals and expectations for the knowledge and skills that will help students succeed. They allow students to understand what is expected of them and to become progressively more proficient in understanding and using mathematics and English language arts. Teachers will be better equipped to know exactly what they must do to help students learn and to establish individualized benchmarks for them.

Will the Common Core State Standards tell teachers how and what to teach?

No. Because the best understanding of what works in the classroom comes from teachers, these standards will establish *what* students need to learn, but they will not dictate *how* teachers should teach. Instead, schools and teachers will decide how best to help students reach the standards.

What will the Common Core State Standards mean for students?

The standards will provide a clear, consistent understanding of what is expected of student learning across the country. Common standards will not prevent different levels of achievement among students, but they will ensure more consistent exposure to materials and learning experiences through curriculum, instruction, teacher preparation, and other supports for student learning. These standards will help give students the knowledge and skills they need to succeed in college and careers.

Do the Common Core State Standards focus on skills and content knowledge?

Yes. The Common Core State Standards recognize that both content and skills are important. They require rigorous content and application of knowledge through higher-order thinking skills. The English language arts standards require certain critical content for all students, including classic myths and stories from around the world, America's founding documents, foundational American literature, and Shakespeare. The remaining crucial decisions about content are left to state and local determination. In addition to content coverage, the Common Core State Standards require that students systematically acquire knowledge of literature and other disciplines through reading, writing, speaking, and listening.

In mathematics, the Common Core State Standards lay a solid foundation in whole numbers, addition, subtraction, multiplication, division, fractions, and decimals. Together, these elements support a student's ability to learn and apply more demanding math concepts and procedures.

The Common Core State Standards require that students develop a depth of understanding and ability to apply English language arts and mathematics to novel situations, as college students and employees regularly do.

Will common assessments be developed?

It will be up to the states: some states plan to come together voluntarily to develop a common assessment system. A state-led consortium on assessment would be grounded in the following principles: allowing for comparison across students, schools, districts, states and nations; creating economies of scale; providing information and supporting more effective teaching and learning; and preparing students for college and careers.

Table of Contents

Common Core State Standards Correlation Chart

Common Core State Standard	Grade 6	*Coach* Lesson(s)
	Domain – Ratios and Proportional Relationships	
Understand ratio concepts and use ratio reasoning to solve problems.		
6.RP.1	Understand the concept of a ratio and use ratio language to describe a ratio relationship between two quantities. *For example, "The ratio of wings to beaks in the bird house at the zoo was 2:1, because for every 2 wings there was 1 beak." "For every vote candidate A received, candidate C received nearly three votes."*	12
6.RP.2	Understand the concept of a unit rate $\frac{a}{b}$ associated with a ratio $a:b$ with $b \neq 0$, and use rate language in the context of a ratio relationship. *For example, "This recipe has a ratio of 3 cups of flour to 4 cups of sugar, so there is* $\frac{3}{4}$ *cup of flour for each cup of sugar." "We paid $75 for 15 hamburgers, which is a rate of $5 per hamburger."*	14
6.RP.3	Use ratio and rate reasoning to solve real-world and mathematical problems, e.g., by reasoning about tables of equivalent ratios, tape diagrams, double number line diagrams, or equations.	
6.RP.3.a	Make tables of equivalent ratios relating quantities with whole-number measurements, find missing values in the tables, and plot the pairs of values on the coordinate plane. Use tables to compare ratios.	13
6.RP.3.b	Solve unit rate problems including those involving unit pricing and constant speed. *For example, if it took 7 hours to mow 4 lawns, then at that rate, how many lawns could be mowed in 35 hours? At what rate were lawns being mowed?*	14
6.RP.3.c	Find a percent of a quantity as a rate per 100 (e.g., 30% of a quantity means $\frac{30}{100}$ times the quantity); solve problems involving finding the whole, given a part and the percent.	15
6.RP.3.d	Use ratio reasoning to convert measurement units; manipulate and transform units appropriately when multiplying or dividing quantities.	16

Common Core State Standard	Grade 6	*Coach* Lesson(s)
Domain – The Number System		
Apply and extend previous understandings of multiplication and division to divide fractions by fractions.		
6.NS.1	Interpret and compute quotients of fractions, and solve word problems involving division of fractions by fractions, e.g., by using visual fraction models and equations to represent the problem. *For example, create a story context for $\left(\frac{2}{3}\right) \div \left(\frac{3}{4}\right)$ and use a visual fraction model to show the quotient; use the relationship between multiplication and division to explain that $\left(\frac{2}{3}\right) \div \left(\frac{3}{4}\right) = \frac{8}{9}$ because $\frac{3}{4}$ of $\frac{8}{9}$ is $\frac{2}{3}$. (In general, $\left(\frac{a}{b}\right) \div \left(\frac{c}{d}\right) = \frac{ad}{bc}$.) How much chocolate will each person get if 3 people share $\frac{1}{2}$ lb of chocolate equally? How many $\frac{3}{4}$-cup servings are in $\frac{2}{3}$ of a cup of yogurt? How wide is a rectangular strip of land with length $\frac{3}{4}$ mi and area $\frac{1}{2}$ square mi?*	9
Compute fluently with multi-digit numbers and find common factors and multiples.		
6.NS.2	Fluently divide multi-digit numbers using the standard algorithm.	2
6.NS.3	Fluently add, subtract, multiply, and divide multi-digit decimals using the standard algorithm for each operation.	7, 8
6.NS.4	Find the greatest common factor of two whole numbers less than or equal to 100 and the least common multiple of two whole numbers less than or equal to 12. Use the distributive property to express a sum of two whole numbers 1–100 with a common factor as a multiple of a sum of two whole numbers with no common factor. *For example, express 36 + 8 as 4(9 + 2).*	1
Apply and extend previous understandings of numbers to the system of rational numbers.		
6.NS.5	Understand that positive and negative numbers are used together to describe quantities having opposite directions or values (e.g., temperature above/below zero, elevation above/below sea level, credits/debits, positive/negative electric charge); use positive and negative numbers to represent quantities in real-world contexts, explaining the meaning of 0 in each situation.	3, 5
6.NS.6	Understand a rational number as a point on the number line. Extend number line diagrams and coordinate axes familiar from previous grades to represent points on the line and in the plane with negative number coordinates.	
6.NS.6.a	Recognize opposite signs of numbers as indicating locations on opposite sides of 0 on the number line; recognize that the opposite of the opposite of a number is the number itself, e.g., $-(-3) = 3$, and that 0 is its own opposite.	3
6.NS.6.b	Understand signs of numbers in ordered pairs as indicating locations in quadrants of the coordinate plane; recognize that when two ordered pairs differ only by signs, the locations of the points are related by reflections across one or both axes.	10, 11
6.NS.6.c	Find and position integers and other rational numbers on a horizontal or vertical number line diagram; find and position pairs of integers and other rational numbers on a coordinate plane.	3, 5, 10

Common Core State Standard	Grade 6	*Coach* Lesson(s)
Domain – The Number System ***(continued)***		
Apply and extend previous understandings of numbers to the system of rational numbers. ***(contined)***		
6.NS.7	Understand ordering and absolute value of rational numbers.	
6.NS.7.a	Interpret statements of inequality as statements about the relative position of two numbers on a number line diagram. *For example, interpret* $-3 > -7$ *as a statement that* -3 *is located to the right of* -7 *on a number line oriented from left to right.*	6
6.NS.7.b	Write, interpret, and explain statements of order for rational numbers in real-world contexts. *For example, write* $-3°C > -7°C$ *to express the fact that* $-3°C$ *is warmer than* $-7°C$.	6
6.NS.7.c	Understand the absolute value of a rational number as its distance from 0 on the number line; interpret absolute value as magnitude for a positive or negative quantity in a real-world situation. *For example, for an account balance of* -30 *dollars, write* $\lvert -30 \rvert = 30$ *to describe the size of the debt in dollars.*	4, 5
6.NS.7.d	Distinguish comparisons of absolute value from statements about order. *For example, recognize that an account balance less than* -30 *dollars represents a debt greater than 30 dollars.*	4, 6
6.NS.8	Solve real-world and mathematical problems by graphing points in all four quadrants of the coordinate plane. Include use of coordinates and absolute value to find distances between points with the same first coordinate or the same second coordinate.	11
Domain – Expressions and Equations		
Apply and extend previous understandings of arithmetic to algebraic expressions.		
6.EE.1	Write and evaluate numerical expressions involving whole-number exponents.	17, 18
6.EE.2	Write, read, and evaluate expressions in which letters stand for numbers.	
6.EE.2.a	Write expressions that record operations with numbers and with letters standing for numbers. *For example, express the calculation "Subtract y from 5" as* $5 - y$.	17
6.EE.2.b	Identify parts of an expression using mathematical terms (sum, term, product, factor, quotient, coefficient); view one or more parts of an expression as a single entity. *For example, describe the expression* $2(8 + 7)$ *as a product of two factors; view* $(8 + 7)$ *as both a single entity and a sum of two terms.*	17
6.EE.2.c	Evaluate expressions at specific values of their variables. Include expressions that arise from formulas used in real-world problems. Perform arithmetic operations, including those involving whole-number exponents, in the conventional order when there are no parentheses to specify a particular order (Order of Operations). *For example, use the formulas* $V = s^3$ *and* $A = 6s^2$ *to find the volume and surface area of a cube with sides of length* $s = \frac{1}{2}$.	18

Common Core State Standard	Grade 6	*Coach* Lesson(s)
Domain – Expressions and Equations *(continued)*		
Apply and extend previous understandings of arithmetic to algebraic expressions. *(continued)*		
6.EE.3	Apply the properties of operations to generate equivalent expressions. *For example, apply the distributive property to the expression $3(2 + x)$ to produce the equivalent expression $6 + 3x$; apply the distributive property to the expression $24x + 18y$ to produce the equivalent expression $6(4x + 3y)$; apply properties of operations to $y + y + y$ to produce the equivalent expression $3y$.*	19
6.EE.4	Identify when two expressions are equivalent (i.e., when the two expressions name the same number regardless of which value is substituted into them). *For example, the expressions $y + y + y$ and $3y$ are equivalent because they name the same number regardless of which number y stands for.*	19
Reason about and solve one-variable equations and inequalities.		
6.EE.5	Understand solving an equation or inequality as a process of answering a question: which values from a specified set, if any, make the equation or inequality true? Use substitution to determine whether a given number in a specified set makes an equation or inequality true.	20, 23
6.EE.6	Use variables to represent numbers and write expressions when solving a real-world or mathematical problem; understand that a variable can represent an unknown number, or, depending on the purpose at hand, any number in a specified set.	17, 22, 23
6.EE.7	Solve real-world and mathematical problems by writing and solving equations of the form $x + p = q$ and $px = q$ for cases in which p, q and x are all nonnegative rational numbers.	20, 22
6.EE.8	Write an inequality of the form $x > c$ or $x < c$ to represent a constraint or condition in a real-world or mathematical problem. Recognize that inequalities of the form $x > c$ or $x < c$ have infinitely many solutions; represent solutions of such inequalities on number line diagrams.	23
Represent and analyze quantitative relationships between dependent and independent variables.		
6.EE.9	Use variables to represent two quantities in a real-world problem that change in relationship to one another; write an equation to express one quantity, thought of as the dependent variable, in terms of the other quantity, thought of as the independent variable. Analyze the relationship between the dependent and independent variables using graphs and tables, and relate these to the equation. *For example, in a problem involving motion at constant speed, list and graph ordered pairs of distances and times, and write the equation $d = 65t$ to represent the relationship between distance and time.*	21, 22
Domain – Geometry		
Solve real.world and mathematical problems involving area, surface area, and volume.		
6.G.1	Find the area of right triangles, other triangles, special quadrilaterals, and polygons by composing into rectangles or decomposing into triangles and other shapes; apply these techniques in the context of solving real-world and mathematical problems.	24–26, 28

Common Core State Standard	Grade 6	*Coach* Lesson(s)
Domain – Geometry *(continued)*		
Solve real-world and mathematical problems involving area, surface area, and volume. *(continued)*		
6.G.2	Find the volume of a right rectangular prism with fractional edge lengths by packing it with unit cubes of the appropriate unit fraction edge lengths, and show that the volume is the same as would be found by multiplying the edge lengths of the prism. Apply the formulas $V = lwh$ and $V = bh$ to find volumes of right rectangular prisms with fractional edge lengths in the context of solving real world and mathematical problems.	31
6.G.3	Draw polygons in the coordinate plane given coordinates for the vertices; use coordinates to find the length of a side joining points with the same first coordinate or the same second coordinate. Apply these techniques in the context of solving real-world and mathematical problems.	27
6.G.4	Represent three-dimensional figures using nets made up of rectangles and triangles, and use the nets to find the surface area of these figures. Apply these techniques in the context of solving real-world and mathematical problems.	29, 30
Domain – Statistics and Probability		
Develop understanding of statistical variability.		
6.SP.1	Recognize a statistical question as one that anticipates variability in the data related to the question and accounts for it in the answers. *For example, "How old am I?" is not a statistical question, but "How old are the students in my school?" is a statistical question because one anticipates variability in students' ages.*	32
6.SP.2	Understand that a set of data collected to answer a statistical question has a distribution which can be described by its center, spread, and overall shape.	32, 33
6.SP.3	Recognize that a measure of center for a numerical data set summarizes all of its values with a single number, while a measure of variation describes how its values vary with a single number.	32, 33
Summarize and describe distributions.		
6.SP.4	Display numerical data in plots on a number line, including dot plots, histograms, and box plots.	34, 36, 37
6.SP.5	Summarize numerical data sets in relation to their context, such as by:	
6.SP.5.a	Reporting the number of observations.	34, 35, 37
6.SP.5.b	Describing the nature of the attribute under investigation, including how it was measured and its units of measurement.	36
6.SP.5.c	Giving quantitative measures of center (median and/or mean) and variability (interquartile range and/or mean absolute deviation), as well as describing any overall pattern and any striking deviations from the overall pattern with reference to the context in which the data were gathered.	32–34, 36, 37
6.SP.5.d	Relating the choice of measures of center and variability to the shape of the data distribution and the context in which the data were gathered.	35

Domain 1

The Number System

Domain 1: Diagnostic Assessment for Lessons 1–11

Domain 1: Cumulative Assessment for Lessons 1–11

Domain 1: Diagnostic Assessment for Lessons 1–11

1. Which situation would you describe with a negative integer?

   A. an airplane flying at an altitude of 30,000 feet

   B. a deposit of $50 into a savings account

   C. a temperature of 30°F

   D. a submarine cruising at 100 feet below sea level

2. Which is the opposite of -17?

   A. 71

   B. 17

   C. -17

   D. -71

3. Which is equivalent to $|-50|$?

   A. -50

   B. -5

   C. 5

   D. 50

4. Which point on the number line represents 0.3?

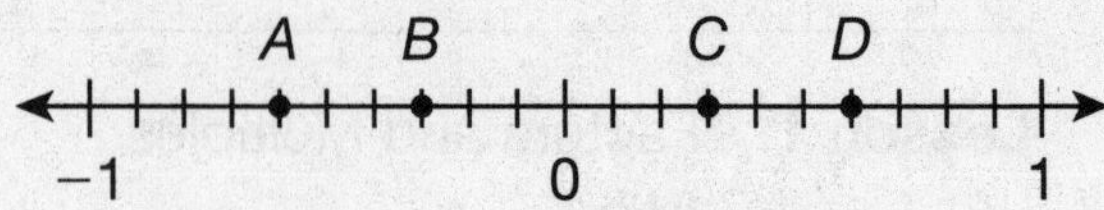

   A. point A

   B. point B

   C. point C

   D. point D

5. Which expression is equivalent to 16 + 56?

   A. $2(14 + 54)$

   B. $4(4 + 12)$

   C. $7(2 + 8)$

   D. $8(2 + 7)$

6. If the numbers below were ordered from least to greatest, which number could you use to replace the ☐?

   $0.23, \frac{2}{5}, \frac{1}{2}, \square, \frac{7}{10}$

   A. 0.45

   B. 0.6

   C. $\frac{7}{8}$

   D. 0.95

7. What is 13,725 ÷ 45?

A. 35

B. 305

C. 350

D. 3,005

8. Which shows how you can check that $\frac{3}{4} \div \frac{5}{6} = \frac{9}{10}$?

A. $\frac{9}{10} \div \frac{5}{6} = \frac{3}{4}$

B. $\frac{9}{10} \div \frac{3}{4} = \frac{5}{6}$

C. $\frac{5}{6} \times \frac{9}{10} = \frac{3}{4}$

D. $\frac{6}{5} \times \frac{9}{10} = \frac{3}{4}$

9. Plot and label point *P* at (4, −3) on the coordinate grid.

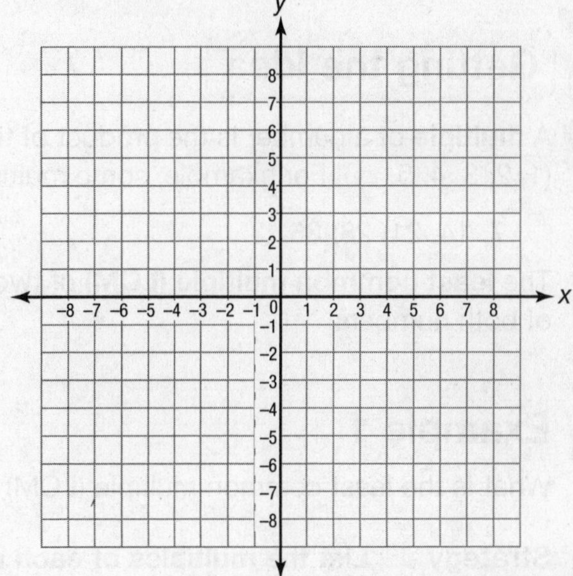

10. Mitch is buying hamburgers and hamburger buns. Hamburgers are sold in packages of 6 each. Hamburger buns are sold in packages of 8 each.

A. What is the least number of hamburgers and hamburger buns Mitch can buy if he wants to have an equal number of each?

____

B. Explain how you found your answer for part A.

____

____

____

Common Core State Standard:
6.NS.4

Factors and Multiples

Getting the Idea

A **multiple** of a number is the product of that number and any of the counting numbers (1, 2, 3, 4, 5, ...). For example, some multiples of 7 are shown below.

7, 14, 21, 28, 35, ...

The **least common multiple (LCM)** of two numbers is the least number that is a multiple of both numbers.

Example 1

What is the least common multiple (LCM) of 8 and 12?

Strategy **List the multiples of each number. Find the least number that is in both lists.**

Step 1 List the first six multiples of 8.

8, 16, 24, 32, 40, 48

Step 2 List the first six multiples of 12.

12, 24, 36, 48, 60, 72

Step 3 What is the least common multiple?

The numbers 24 and 48 appear in both lists, but 24 is the *least* of those multiples.

Solution **The least common multiple (LCM) of 8 and 12 is 24.**

You can use what you know about the least common multiple to solve a real-world problem.

Example 2

The students in the drama club can be divided into 3 equal groups or 5 equal groups, with no students left over. What is the least number of students that could be in the drama club?

Strategy **Use what you know about the least common multiple.**

Step 1 How can the LCM help you solve this problem?

Since the drama club can be divided into equal groups of 3 or 5, the number of students in the drama club must be a multiple of both 3 and 5.

To find the least possible number of students in the drama club, find the LCM of 3 and 5.

Step 2 List the first six multiples of 3 and 5.

Multiples of 3: 3, 6, 9, 12, 15, 18

Multiples of 5: 5, 10, 15, 20, 25, 30

Step 3 Find the least number that is common to both lists.

The least common multiple is 15.

Solution **The least possible number of students in the drama club is 15.**

The **factors** of a number are the counting numbers that can be multiplied to get that number. (A factor of a number is also any counting number that divides that number evenly.) For example:

$1 \times 10 = 10$, so 1 and 10 are factors of 10.

$2 \times 5 = 10$, so 2 and 5 are also factors of 10.

The number 10 has four factors: 1, 2, 5, and 10.

The greatest common factor (GCF) of two numbers is the greatest number that is a factor of both numbers.

Example 3

What is the greatest common factor (GCF) of 20 and 50?

Strategy **List the factors of each number. Find the greatest number that is in both lists.**

Step 1 List the factors of 20.

1, 2, 4, 5, 10, 20

Step 2 List the factors of 50.

1, 2, 5, 10, 25, 50

Step 3 What is the greatest common factor?

1, 2, 5, and 10 appear in both lists.

The *greatest* of those common factors is 10.

Solution **The greatest common factor (GCF) of 20 and 50 is 10.**

Numbers can be expressed in different ways. For example, the number 10 can be written as the sum 4 + 6. You can use the **distributive property** to write this sum in another way.

Distributive Property

When you multiply the sum of two numbers by another number, you can multiply each addend by the number, and then add the products. The property also applies to subtraction.

$$a(b + c) = ab + ac$$
$$a(b - c) = ab - ac$$

Using the distributive property: $2(2 + 3) = (2 \times 2) + (2 \times 3)$
$= 4 + 6$
$= 10$

So, you can write 10 as 4 + 6 and as 2(2 + 3). Notice that the numbers in the sum (2 + 3) have no common factors.

Example 4

The number 99 can be expressed as the sum 45 + 54. Use the distributive property to rewrite that sum as a multiple of a sum whose addends have no common factors.

Strategy **Find the greatest common factor of 45 and 54. Use the distributive property to rewrite the sum.**

Step 1 Find the greatest common factor of 45 and 54.

Factors of 45: 1, 3, 5, 9, 15, and 45

Factors of 54: 1, 2, 3, 6, 9, 18, 27, and 54

The common factors are 1, 3, and 9. So, the GCF is 9.

Step 2 Use the GCF and the distributive property to rewrite the sum.

$45 = 5 \times 9$

$54 = 6 \times 9$

$45 + 54 = 9(5 + 6)$

Step 3 Check that the addends in the sum have no common factors, other than 1.

Factors of 5: 1, 5

Factors of 6: 1, 2, 3, 6

The numbers 5 and 6 have no common factors, other than 1.

Solution **The sum 45 + 54 can be expressed as 9(5 + 6).**

Coached Example

Two P.E. classes are participating in Field Day. One class has 24 students. The other class has 27 students. The P.E. teachers want to divide the students in each class into the largest possible equal groups, with no students left over. If all the groups have the same number of students, how many students are in each group?

If the classes are divided into equal groups, the number of students in each group must be a ______________ of 24 and 27.

List the factors of 24: 1, _____, _____, _____, _____, _____, _____, 24

List the factors of 27: 1, _____, _____, 27

What are the common factors of 24 and 27? ______________

What is the greatest common factor of 24 and 27? _____

If all the groups have the same number of students, there will be _____ students in each group.

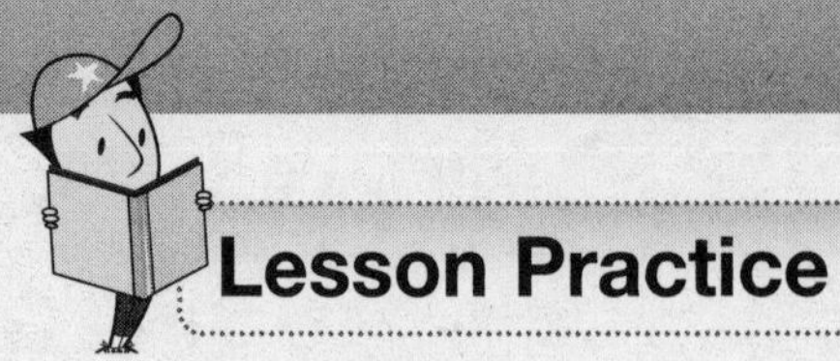

Lesson Practice

Choose the correct answer.

1. What is the greatest common factor (GCF) of 25 and 35?

   A. 1
   B. 5
   C. 7
   D. 25

2. What is the least common multiple (LCM) of 2 and 6?

   A. 12
   B. 10
   C. 8
   D. 6

3. What is the GCF of 24 and 36?

   A. 12
   B. 9
   C. 6
   D. 3

4. What is the LCM of 6 and 10?

   A. 12
   B. 20
   C. 30
   D. 60

5. Which of the following is equivalent to $33 + 77$?

   A. $3(11 + 7)$
   B. $(3 \times 11) \times (7 \times 11)$
   C. $7(3 + 11)$
   D. $11(3 + 7)$

6. The number 108 can be expressed as the sum $100 + 8$. Which shows how to use the distributive property to rewrite that sum as a multiple of a sum whose addends have no common factors?

   A. $2(50 + 4)$
   B. $4(25 + 2)$
   C. $5(20 + 1)$
   D. $8(12 + 1)$

7. Ms. Madison directs two choruses. One chorus has 28 students. The other chorus has 36 students. For rehearsals, she wants to divide each chorus into the largest possible equal groups, with no students left over. How many students will be in each group?

A. 2

B. 4

C. 9

D. 12

8. Two airport shuttle trains leave the main station at the same time. Shuttle A returns to the station every 8 minutes. Shuttle B returns to the station every 10 minutes. In how many minutes will Shuttles A and B leave the station together for the second time?

A. 10 minutes

B. 18 minutes

C. 40 minutes

D. 80 minutes

9. Evan bought two plants. He decided to water his first plant every 3 days and his second plant every 4 days.

A. If he watered both plants on June 1, how many days passed before he watered both plants on the same day again? Show or explain your work.

B. On June 25, Evan decided that he was not watering his first plant frequently enough. He started watering his first plant every 2 days. He continued to water his second plant every 4 days. If he watered both plants that day, how many days passed before he watered both plants on the same day again? Show or explain your work.

Common Core State Standard:
6.NS.2

Divide Whole Numbers

Getting the Idea

In division, the number that is divided is the **dividend**. The number that divides the dividend is the **divisor**. The answer to a division problem is the **quotient**. Some division problems will have a remainder. A **remainder** is a counting number that is left over when two counting numbers are divided. A remainder is always less than the divisor.

To divide by a 2-digit number, you may need to estimate first to help you find the quotient.

Example 1

What is 64,015 ÷ 74?

Strategy **Estimate the first digit in the quotient and work from there.**

Step 1 Decide where to place the first digit in the quotient.

$$74\overline{)64{,}015}$$

The first digit of the quotient will be in the hundreds place.

Step 2 Divide 640 by 74.

You know that 8 × 70 = 560 and 9 × 70 = 630, so try 8 first.

```
      8
74)64,015
  −59 2      ← Multiply: 8 × 74 = 592
    4 8      ← Subtract: 640 − 592 = 48
```

Step 3 Bring down the 1 and divide.

You know that 6 × 70 = 420, so try 6.

```
      86
74)64,015
  −59 2
    4 81
   −4 44     ← Multiply: 6 × 74 = 444
      37     ← Subtract: 481 − 444 = 37
```

Step 4 Bring down the 5 and divide.

You know that 5 × 70 = 350, so try 5.

```
     865
74)64,015
  −59 2
    4 81
   −4 44
      375
     −370   ← Multiply: 5 × 74 = 370
        5   ← Subtract: 375 − 370 = 5
              The remainder is 5.
```

Solution **64,015 ÷ 74 = 865 R5**

Example 2

There are 1,288 seats in an auditorium. Each of the 23 rows in the auditorium has the same number of seats. How many seats are in each row?

Strategy **Divide to find the solution.**

Step 1 Decide where to place the first digit in the quotient.

```
23)1,288
```

The first digit of the quotient will be in the tens place.

Step 2 Divide 128 by 23.

```
       5
23)1,288
  −1 15     ← Multiply: 5 × 23 = 115
     13     ← Subtract: 128 − 115 = 13
```

Step 3 Bring down the 8 and divide.

```
      56
23)1,288
  −1 15
     138
    −138    ← Multiply: 6 × 23 = 138
       0    ← Subtract: 138 − 138 = 0
```

Solution **There are 56 seats in each row of the auditorium.**

Example 3

Mindy's annual salary as a physical therapist is $59,796. How much does Mindy earn per month?

Strategy **Divide each place, going from left to right.**

Step 1 There are 12 months in a year, so the divisor is 12.

```
12)59,796
```

The first digit of the quotient will be in the thousands place.

Step 2 Divide 59 by 12.

```
     4
12)59,796
   -48         ← Multiply: 4 × 12 = 48
    11         ← Subtract: 59 − 48 = 11
```

Step 3 Bring down the 7. Divide.

```
     4 9
12)59,796
   -48
    11 7
   -10 8       ← Multiply: 9 × 12 = 108
       9       ← Subtract: 117 − 108 = 9
```

Step 4 Bring down the 9. Divide.

```
     4 98
12)59,796
   -48
    11 7
   -10 8
       99
      -96      ← Multiply: 8 × 12 = 96
        3      ← Subtract: 99 − 96 = 3
```

Step 5 Bring down the 6. Divide.

```
      4 983
 12)59,796
   -48
    11 7
   -10 8
       99
      -96
       36
      -36   ← Multiply: 3 × 12 = 36
        0   ← Subtract: 36 − 36 = 0
```

Solution **Mindy earns $4,983 per month.**

Coached Example

Divide: 32)89,824

```
     2 _ 0 _
 32)8 9,8 2 4
   -6 4          ← Multiply: 2 × 32 = ____
    _ _ 8
   − _ _ _       ← Multiply: __ × 32 = ____
        2 2
   −      0      ← Multiply: ___ × 32 = ____
       _ _ 4
     − _ _ _     ← Multiply: ___ × 32 = ____
           0
```

Check your answer. Multiply the quotient and the divisor.

__________ × 32 = __________

89,824 ÷ 32 = __________

Lesson Practice

Choose the correct answer.

1. What is 2,520 ÷ 36?

A. 7
B. 70
C. 170
D. 210

2. What is 6,854 ÷ 17?

A. 40 R3
B. 403
C. 403 R3
D. 430

3. What is 11,362 ÷ 46?

A. 247
B. 248
C. 252
D. 253

4. What is 72,450 ÷ 25?

A. 2,888
B. 2,892
C. 2,898
D. 2,902

5. Mr. and Mrs. Chin flew from New York to Tokyo, which is a distance of 6,375 miles. If it took the plane 15 hours to fly from New York to Tokyo, what was the plane's average speed per hour?

A. 415 miles per hour
B. 425 miles per hour
C. 435 miles per hour
D. 475 miles per hour

6. Eggs are packed 12 to a carton. There are 7,260 eggs to be put in cartons. How many cartons are needed for the eggs?

A. 65
B. 605
C. 625
D. 650

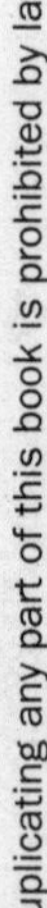

7. Ticket sales for a concert totaled $89,200. Tickets for the concert cost $16 each. How many tickets were sold?

A. 557
B. 575
C. 5,570
D. 5,575

8. Homer's annual salary is $74,308. If he works all 52 weeks of the year, how much is he paid each week?

A. $1,249
B. $1,294
C. $1,429
D. $1,439

9. An airport has 24 gates. One month, 43,776 passengers left through the gates.

A. What was the average number of passengers that left through each gate?

____

B. Explain how you knew where to place the first digit in the quotient in Part A.

____

____

____

Common Core State Standards:
6.NS.5, 6.NS.6.a, 6.NS.6.c

Integers

Getting the Idea

Integers include the counting numbers (1, 2, 3, ...), their opposites (−1, −2, −3, ...), and zero. The number line below shows the integers from −5 to 5. **Negative integers** have values less than zero, so they are to the left of zero on the number line. **Positive integers** have values greater than zero, so they are to the right of zero on the number line. Zero is neither negative nor positive.

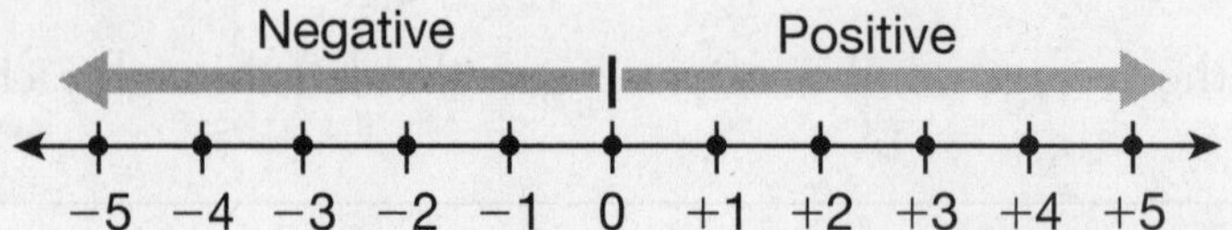

You can use integers to describe opposite situations. Here are some uses for integers:

Positive integers

- A bank deposit (adding money to an account)
- An elevation above sea level
- A rise in temperature

Negative integers

- A bank withdrawal (taking money out of an account)
- An elevation below sea level
- A drop in temperature

Example 1

A bird is flying 25 feet above sea level and a fish is swimming 10 feet below sea level. Use integers to represent the elevation of the fish and the bird.

Strategy **Use an integer to describe each situation.**

What elevation would the number 0 represent?

Zero represents sea level, or the surface of the water.

Step 2 Find a signed number for the elevation of the bird.

The bird is above sea level, so use a positive number (a number greater than 0).

+25 or simply 25

Step 3 Find a signed number for the elevation of the fish.

The fish is below sea level, so use a negative number (a number less than 0).

-10

Solution **The elevation of the bird is 25 feet. The elevation of the fish is −10 feet.**

You can show negative integers by extending to the left a number line that shows the numbers (0, 1, 2, 3, . . .). Number lines showing positive and negative integers can be either horizontal or vertical, such as a thermometer.

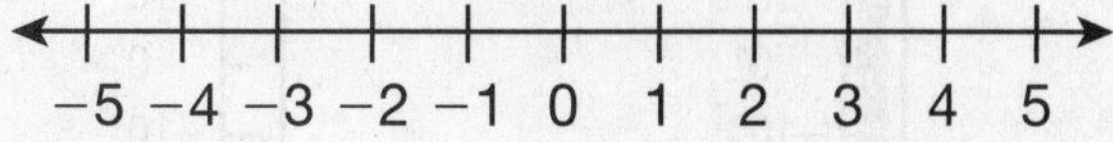

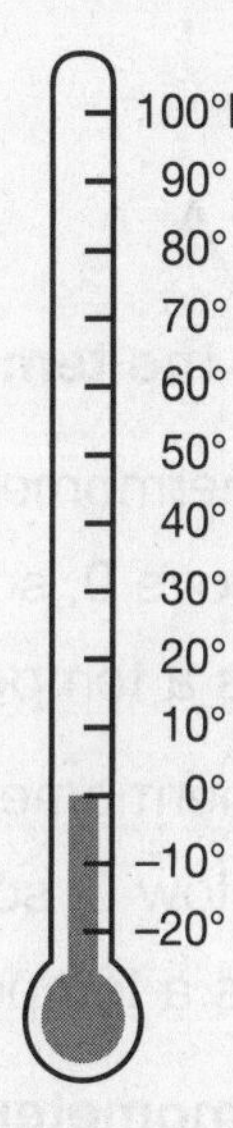

Example 2

What temperatures are indicated on the Fahrenheit thermometers below?

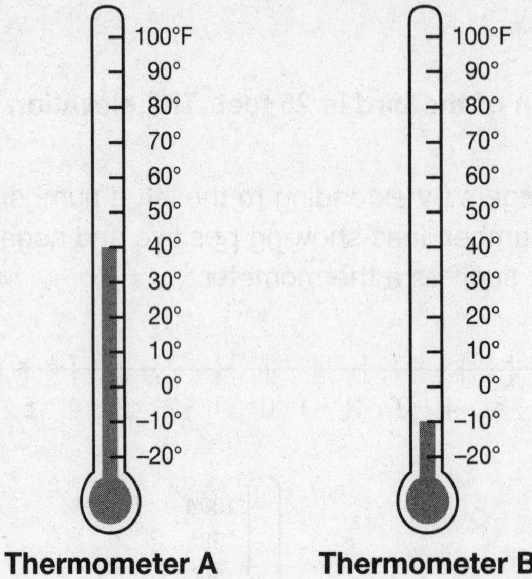

Strategy **Use integers to describe the temperature.**

Step 1 Find the temperature on thermometer A.

The temperature is above 0, so use a positive number.

Thermometer A shows a temperature of 40° F.

Step 2 Find the temperature on thermometer B.

The temperature is below 0, so use a negative number.

Thermometer B shows a temperature of −10°F.

Solution **The temperature on thermometer A is 40°F. The temperature on thermometer B is −10°F.**

Integers that are the same distance from 0 on a number line are called **opposites**. For example, 5 and −5 are opposites of each other. They are each the same distance from 0 on a number line, as shown below.

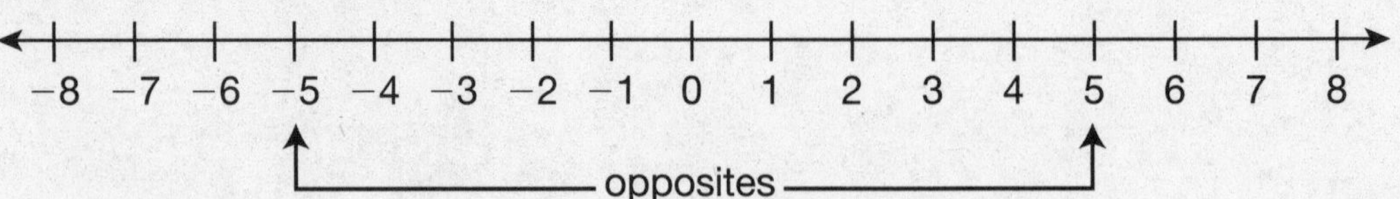

So, the opposite of 5 is −5, and the opposite of −5, written as −(−5), is 5. The opposite of 0 is 0.

Example 3

Find the opposites of 6 and of −2.

Strategy **Use a number line.**

Step 1 Plot a point for 6 on a number line.

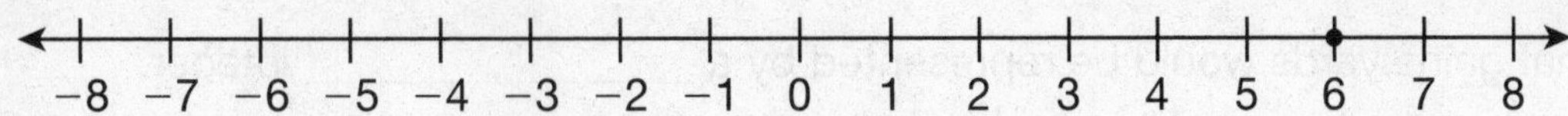

Step 2 Find the integer that is the same distance from 0 in the opposite direction.

The integer 6 is 6 units to the right of 0.

Count 6 units to the left of 0. Plot a point.

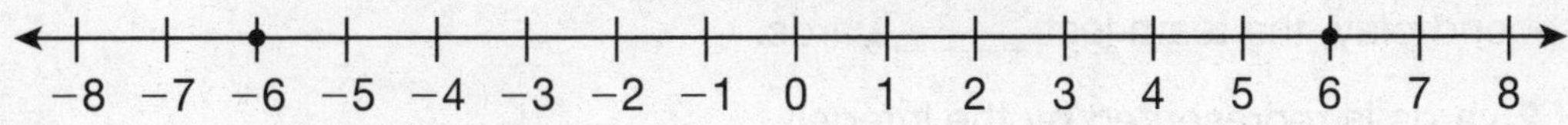

The opposite of 6 is −6.

Step 3 Plot a point for −2 on a number line.

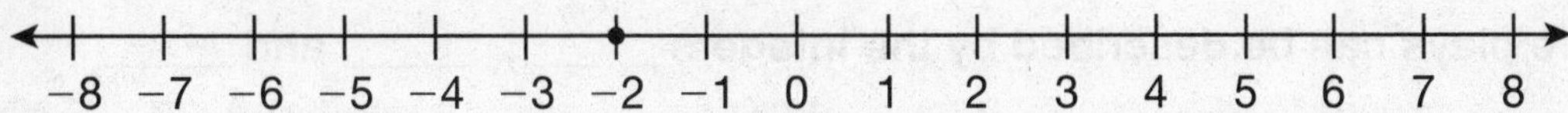

Step 4 Find the integer that is the same distance from 0 in the opposite direction.

The integer −2 is 2 units to the left of 0.

Count 2 units to the right of 0. Plot a point.

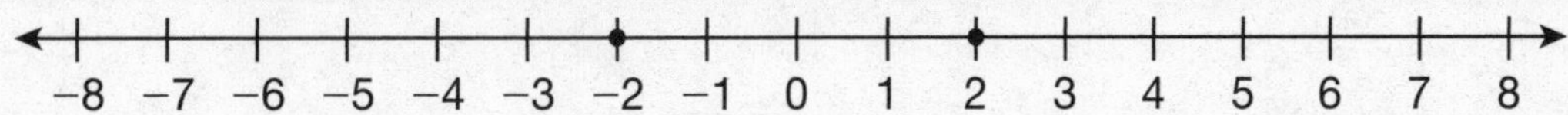

The opposite of −2 is 2.

Solution **The opposite of 6 is −6. The opposite of −2 is 2.**

Coached Example

On its first play of the game, a football team gained 6 yards. On its next two plays, the team lost 2 yards and then gained 7 yards. Use integers to describe these three plays.

What integer represents a play in which the team neither gains yards nor loses yards? ____

A play that gains yards would be represented by a ______________ integer.

A play that loses yards would be represented by a ______________ integer.

On the first play, the team gained ______ yards.

A gain of 6 yards is represented by the integer ______.

On the second play, the team lost ______ yards.

A loss of 2 yards is represented by the integer ______.

On the third play, the team gained ______ yards.

A gain of 7 yards is represented by the integer ______.

The three plays can be described by the integers ______, ______, and ______.

Lesson Practice

Choose the correct answer.

Use the number line for questions 1 and 2.

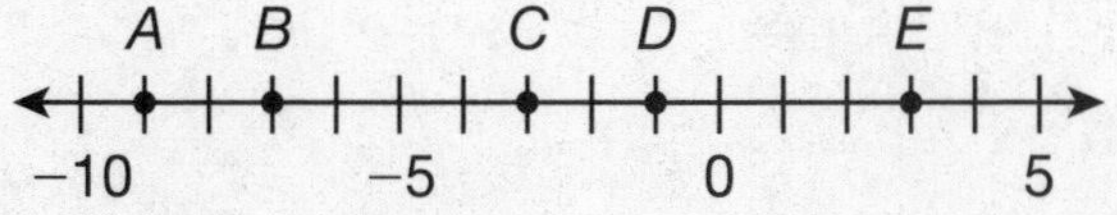

1. Which point on the number line is represented by -7?
   - **A.** point *A*
   - **B.** point *B*
   - **C.** point *C*
   - **D.** point *D*

2. Which integer is represented by point *E*?
   - **A.** -3
   - **B.** -2
   - **C.** 2
   - **D.** 3

3. Which situation can be represented by the integer 10?
   - **A.** a temperature drop of 10°F
   - **B.** 10 seconds before takeoff
   - **C.** a growth of 10 centimeters
   - **D.** a fall of 10 feet

4. Which situation would you represent with a negative integer?
   - **A.** a mountain climber descending a mountain
   - **B.** a price increase
   - **C.** a person winning a sum of money
   - **D.** an elevator going from the 2nd floor to the 5th floor

5. New Orleans has an elevation of 7 feet below sea level. How is that elevation, in feet, represented as an integer?
   - **A.** -17
   - **B.** -7
   - **C.** 7
   - **D.** 177

6. What is the opposite of -13?
   - **A.** 31
   - **B.** 13
   - **C.** -3
   - **D.** -31

7. What is the opposite of 40?

A. −40
B. −4
C. 4
D. 40

8. What is another way to write −(−9)?

A. −99
B. −9
C. 0
D. 9

9. Use the number line below.

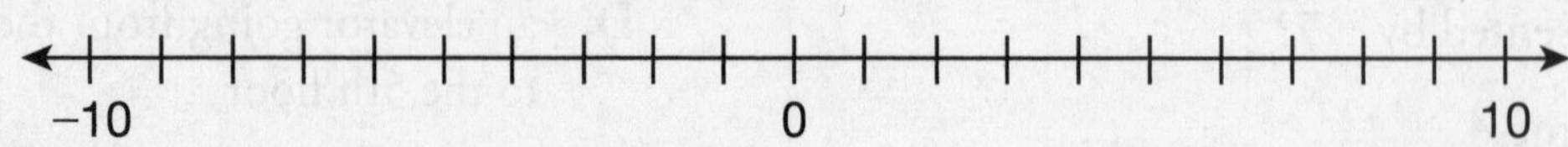

A. Plot and label point *J* at −8, point *K* at 6, point *L* at −1, and point *M* at 4.

B. Point *N* is the opposite of point *J*. Plot and label point *N* on the number line above. Explain how you found the opposite of point *J*.

____

____

____

Common Core State Standards: 6.NS.7.c, 6.NS.7.d

Absolute Value

Getting the Idea

The **absolute value** of a number is its distance from 0 on a number line. Since a distance must be either a positive number or zero, the absolute value of a number is always a positive number or zero. The absolute value of a number x is written as $|x|$.

The integers −4 and 4 are opposites. You can use the number line below to see that each number is the same distance from 0. So, $|-4| = 4$ and $|4| = 4$.

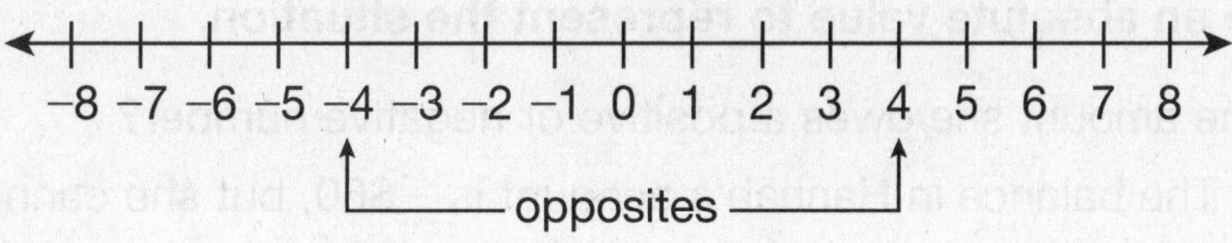

Example 1

Find the value of $|-7|$.

Strategy **Use a number line.**

Step 1 Plot a point for −7 on a number line.

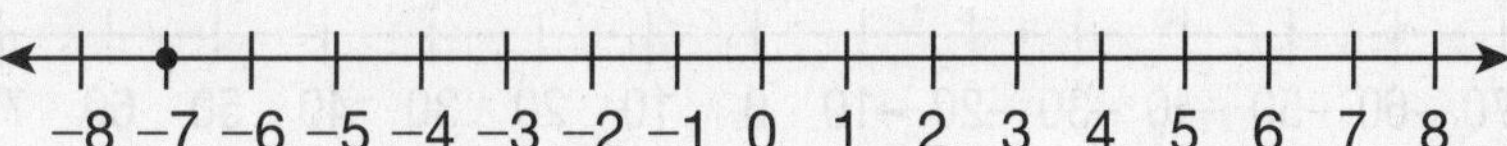

Step 2 Count the number of units from −7 to 0.

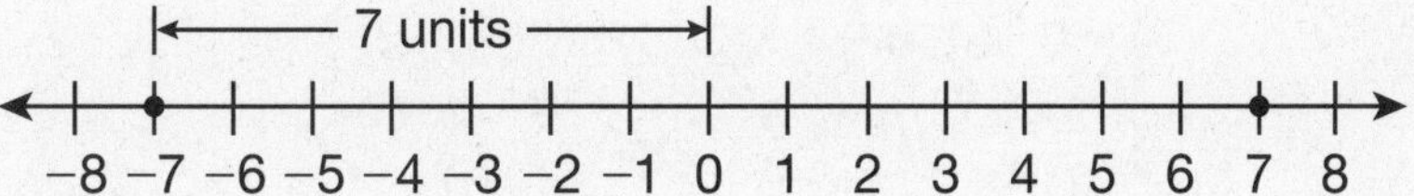

The distance is 7 units.

$|-7| = 7$

Solution $|-7| = 7$

You can use absolute values to represent and help you understand real-world situations.

For example, if a diver is 20 meters below the ocean's surface, that depth, in meters, can be shown as -20 meters. But the distance the diver would have to swim to get to the surface of the water cannot be represented by a negative number. You can use absolute value instead. The diver must swim $|-20|$ meters, or 20 meters, to reach the surface.

Example 2

Hannah wrote a check for more money than she has in her bank account. The balance in her account is now $-\$60$. How much does Hannah owe the bank, in dollars?

Strategy **Use an absolute value to represent the situation.**

Step 1 Is the amount she owes a positive or negative number?

The balance in Hannah's account is $-\$60$, but she cannot owe the bank a negative amount of money.

The amount Hannah owes must be shown as a positive number.

Step 2 Use an absolute value.

The amount she owes, in dollars, is $|-60|$, or 60.

The number line below shows that Hannah owes the bank $60.

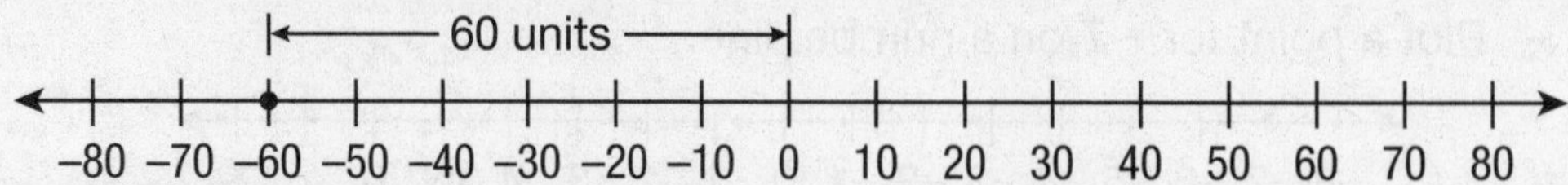

Solution **Hannah owes the bank $60.**

Absolute values can also help you understand situations in which an exact number is not known.

Example 3

A team of mountaineers has climbed to the summit of Mount Everest. The temperature at the summit is less than -15°F. Describe how many degrees Fahrenheit below 0°F the temperature is.

Strategy **Use an absolute value to represent and understand the situation.**

Step 1 Is the number of degrees below 0°F a positive or negative number?

An actual temperature may be negative, but the number of degrees Fahrenheit below 0°F must be a positive number.

Step 2 Use a number line to represent the situation.

The temperature is *less than* -15°F.

On a number line, a number less than -15 is to the left of -15.

The arrow below shows all the numbers less than -15.

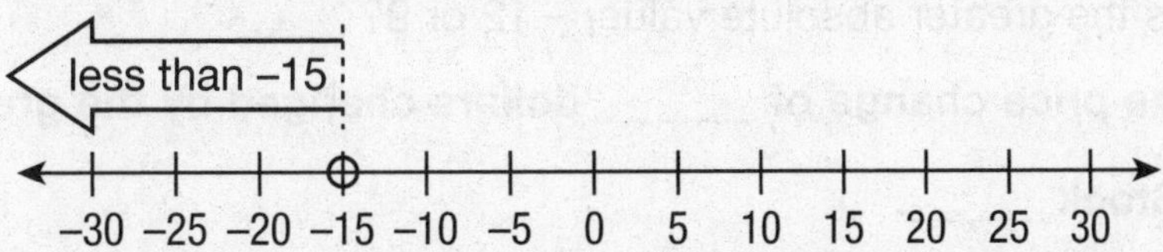

Step 3 Use absolute value to describe the number of degrees Fahrenheit below 0°F.

$|-15| = 15$

less than –15 ←15 units→

–30 –25 –20 –15 –10 –5 0 5 10 15 20 25 30

All the numbers *less than* -15 are *more than* 15 units from 0.

So, if the temperature is *less than* -15°F, it is *more than* 15°F below 0°F.

Solution **The temperature at the summit is more than 15°F below 0°F.**

Coached Example

Yesterday, Marcus bought two different stocks, A and B, each at the same price. From yesterday to today, the change in the price of Stock A was −\$12, and the change in the price of Stock B was \$9. From yesterday to today, which stock's price changed by the greatest amount?

The price change with the greatest __________ __________ is the greatest change.

On the number line below, plot points for −12 and 9.

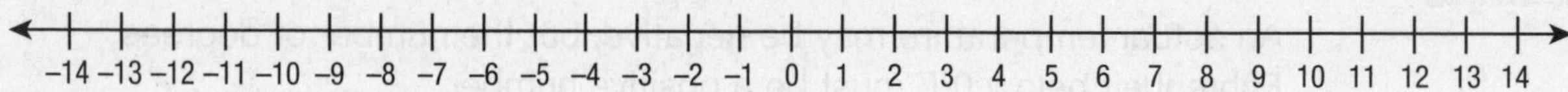

Count the units from each integer to 0 to determine its absolute value.

$|-12| =$ ______

$|9| =$ ______

Which number has the greater absolute value, −12 or 9? ______

The stock with the price change of ______ dollars changed by the greatest amount.

That stock was Stock ______.

Lesson Practice

Choose the correct answer.

1. Which point represents a number with an absolute value of 1?

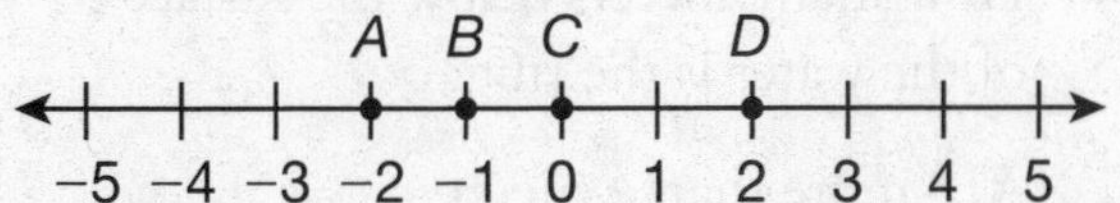

   A. point *A*
   B. point *B*
   C. point *C*
   D. point *D*

2. Which statement best describes $|-72|$?

   A. the distance from −72 to 72 on a number line
   B. the distance from −7 to −2 on a number line
   C. the distance from −7 to 2 on a number line
   D. the distance from −72 to 0 on a number line

3. A scientist stores liquid nitrogen at a temperature of −331°F. Exactly how many degrees Fahrenheit below 0°F is the liquid nitrogen?

   A. 331
   B. 31
   C. 0
   D. −331

4. Which point or points on the number line represent numbers with absolute values of 4?

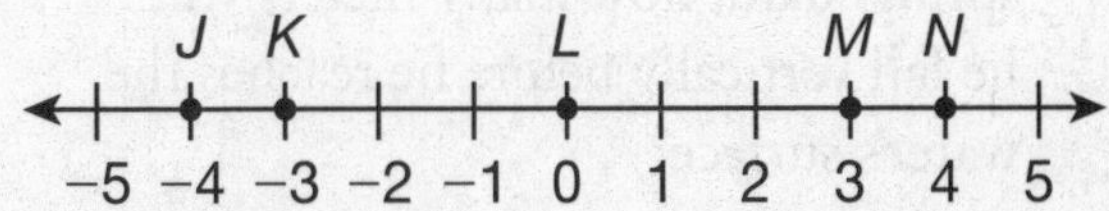

   A. points *J* and *N*
   B. points *K* and *M*
   C. points *K* and *N*
   D. points *L* and *N*

5. Lorraine's checking account has a balance of less than −\$200. Which statement is true about how much Lorraine owes the bank?

   A. Lorraine owes exactly \$200.
   B. Lorraine owes exactly −\$200.
   C. Lorraine owes less than \$200.
   D. Lorraine owes more than \$200.

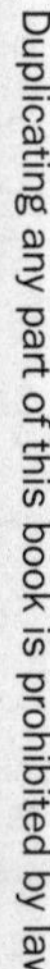

Use the information for questions 6 and 7.

Max is a diver. He uses positive numbers to represent elevations above the water's surface and negative numbers to represent elevations below the water's surface. Max is standing on a springboard. He represents his location as 3 meters. He lets a ring drop to the bottom of the pool. He represents its location at the bottom of the pool as −4 meters.

6. If Max dives into the pool from the springboard, how many meters will he fall vertically before he reaches the water's surface?

A. 4 meters
B. 3 meters
C. −3 meters
D. −4 meters

7. How many meters below the surface of the water is the ring?

A. more than 4 meters
B. less than 4 meters
C. exactly 4 meters
D. exactly 3 meters

8. A football coach recorded the results of his team's first 4 plays in its last game. The table below shows his data.

Football Plays

Play	Number of Yards
1	8
2	−2
3	5
4	−7

A. During which play did the team lose the fewest yards? Use what you know about absolute value and the number line below to explain how you determined your answer.

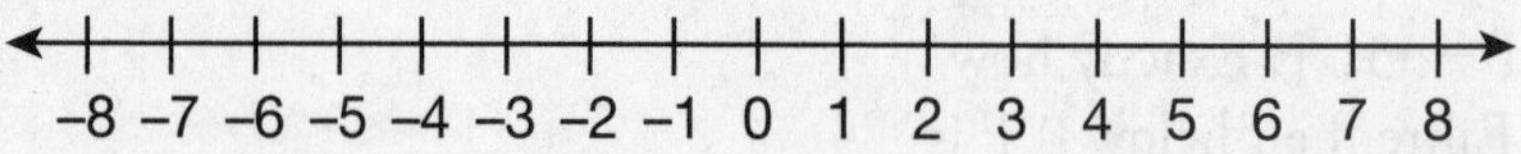

B. During which play did the team gain the most yards? Use what you know about absolute value and the number line above to explain your answer.

Common Core State Standards: 6.NS.5, 6.NS.6.c, 6.NS.7.c

Rational Numbers

Getting the Idea

A **rational number** is a number that can be expressed as the ratio of two integers in the form $\frac{a}{b}$, where b is not equal to 0. A rational number can be positive or negative. The set of rational numbers includes integers, fractions, mixed numbers, percents, terminating decimals, and repeating decimals. Some examples of rational numbers are shown below.

8% $\quad \frac{4}{5} \quad$ 0.35 $\quad 1\frac{3}{8} \quad -7 \quad 1.\overline{6}$

Fractions and decimals have opposites, just as integers do. For example, $\frac{5}{8}$ and $-\frac{5}{8}$ are opposites, and so are −3.25 and 3.25.

You can see rational numbers in many real-world situations, such as a sheet of paper that is $8\frac{1}{2}$ inches wide or a dog that weighs 29.51 kilograms.

Example 1

Explain why 4, $\frac{2}{3}$, and 0.9 are rational numbers.

Strategy **Express the numbers in the form $\frac{a}{b}$.**

Step 1 Show that 4 is a rational number.

$4 = \frac{4}{1}$, which is in the form $\frac{a}{b}$.

Step 2 Show that $\frac{2}{3}$ is a rational number.

$\frac{2}{3}$ is in the form $\frac{a}{b}$.

Step 3 Show that 0.9 is a rational number.

$0.9 = \frac{9}{10}$, which is in the form $\frac{a}{b}$.

Solution **The numbers 4, $\frac{2}{3}$, and 0.9 are rational numbers because each can be written in the form $\frac{a}{b}$.**

Remember that the absolute value of a number is the distance of that number from 0 on a number line. You can use absolute value to help you locate a rational number on a number line.

Example 2

Plot the rational numbers $\frac{3}{5}$ and $-\frac{2}{5}$ on the number line shown below.

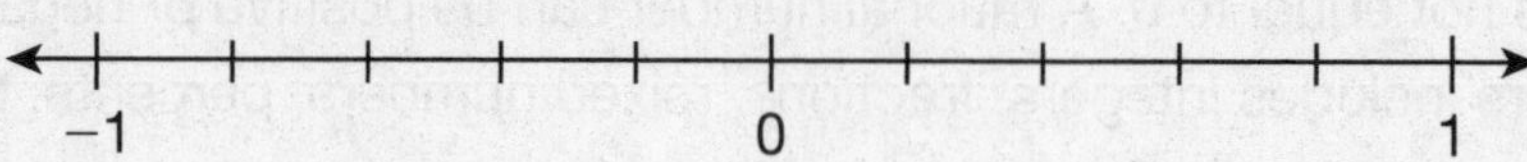

Strategy **Identify what fractional units the number line is divided into. Then use absolute value to plot each point.**

Step 1 Determine what each mark on the number line stands for.

There are 5 spaces between 0 and 1.

The number line is divided into fifths. Each mark stands for $\frac{1}{5}$.

Step 2 Find the absolute value of $\frac{3}{5}$.

$\left|\frac{3}{5}\right| = \frac{3}{5}$ or 3 fifths

Step 3 Plot $\frac{3}{5}$ on the number line.

$\frac{3}{5}$ is three units away from 0 on this number line divided into fifths.

Since $\frac{3}{5}$ is positive, count 3 units to the right of 0. Plot the point.

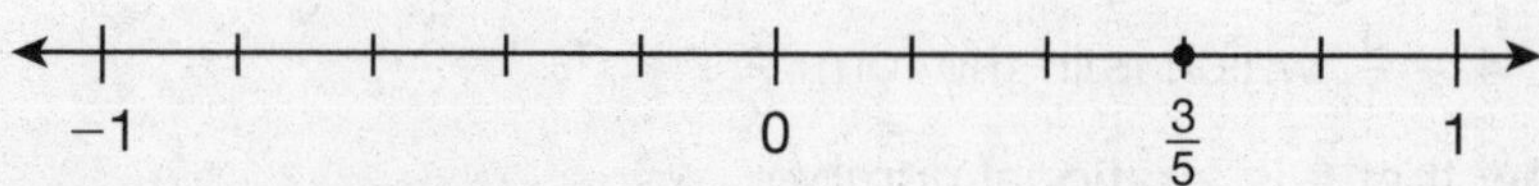

Step 4 Find the absolute value of $-\frac{2}{5}$.

$\left|-\frac{2}{5}\right| = \frac{2}{5}$ or 2 fifths

Step 5 Plot $-\frac{2}{5}$ on the number line.

$-\frac{2}{5}$ is two units away from 0 on this number line divided into fifths.

Since $-\frac{2}{5}$ is negative, count 2 units to the left of 0. Plot the point.

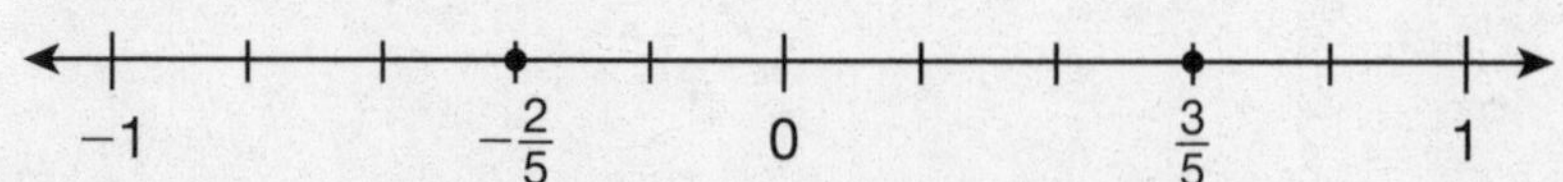

Solution **The rational numbers $\frac{3}{5}$ and $-\frac{2}{5}$ are shown on the number line in Step 5 above.**

Example 3

The floor of the valley in which Griffin lives is $7\frac{1}{2}$ feet below sea level. Write that elevation as a rational number. Then plot a point for it on the number line below.

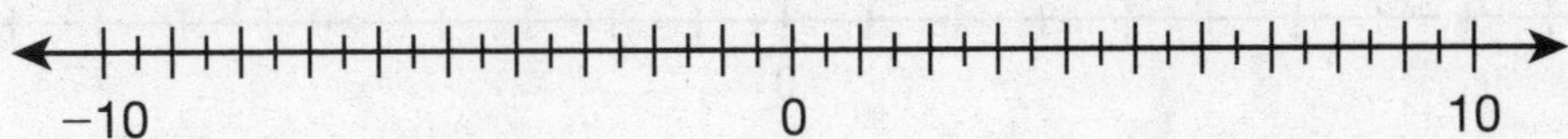

Strategy **Write a rational number representing $7\frac{1}{2}$ feet below sea level. Then identify the value of each mark on the number line.**

Step 1 Determine whether the elevation is a positive or negative number.

In this case, the number 0 represents sea level.

The valley floor is below sea level, so the number will be less than 0, or negative.

The rational number $-7\frac{1}{2}$ represents the elevation.

Step 2 Determine what each mark on the number line stands for.

There are 20 spaces between 0 and 10 and 20 spaces between 0 and −10.

Each mark between integers stands for $\frac{1}{2}$.

Step 3 Find −7 on the number line.

Since −7 is a negative number, it will be to the left of 0 on the number line.

Step 4 Plot $-7\frac{1}{2}$ on the number line.

$\left|-7\frac{1}{2}\right| = 7\frac{1}{2}$ $\qquad |-7| = 7$

$7\frac{1}{2} > 7$, so $-7\frac{1}{2}$ will be farther from 0 than −7.

Plot a point for $-7\frac{1}{2}$ at the mark to the left of −7.

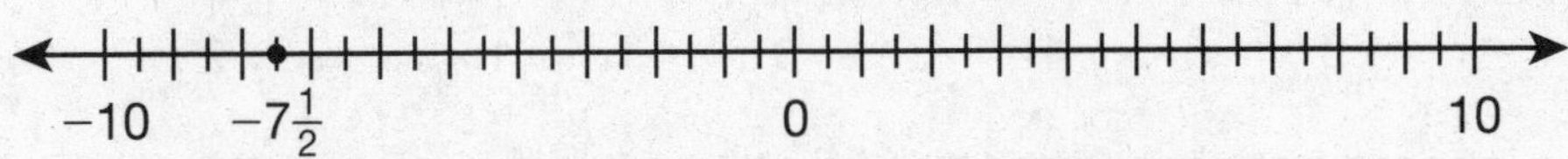

Solution **The rational number $-7\frac{1}{2}$ represents the elevation.**

The location of $-7\frac{1}{2}$ on a number line is shown in Step 4 above.

Example 4

What decimals do points P and Q represent on the number line shown?

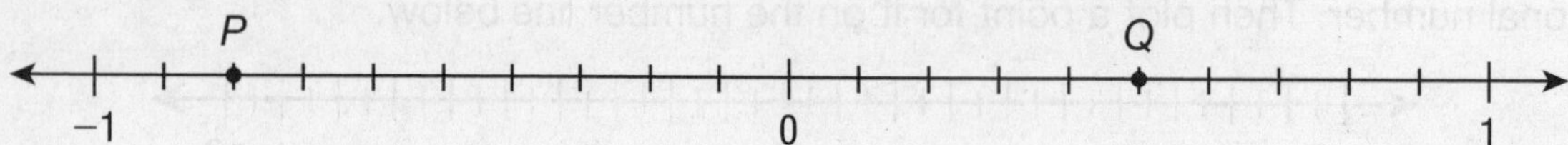

Strategy **Identify what units the number line is divided into.**

Step 1 Determine what each mark on the number line stands for.

There are 10 spaces between 0 and 1.

The number line is divided into tenths. Each mark stands for 0.1.

Step 2 Find the number of tenths that points P and Q represent.

Count from 0. It may help to label the marks as shown below.

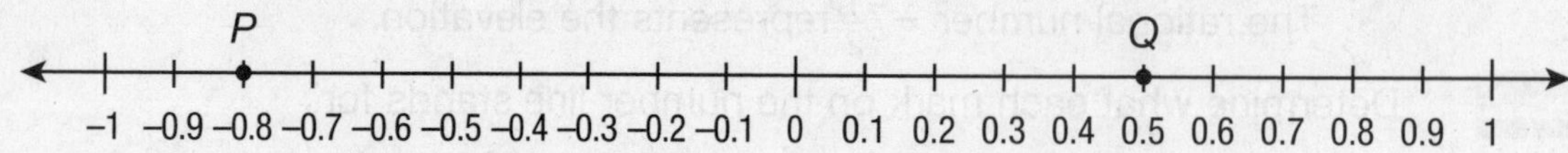

Point P is at −0.8 and point Q represents 0.5.

Solution **Point P represents −0.8 and point Q represents 0.5.**

Coached Example

The number line below shows points for several rational numbers. Which points represent $-\frac{1}{5}$, $\frac{4}{5}$, $2\frac{2}{5}$, and $-1\frac{3}{5}$?

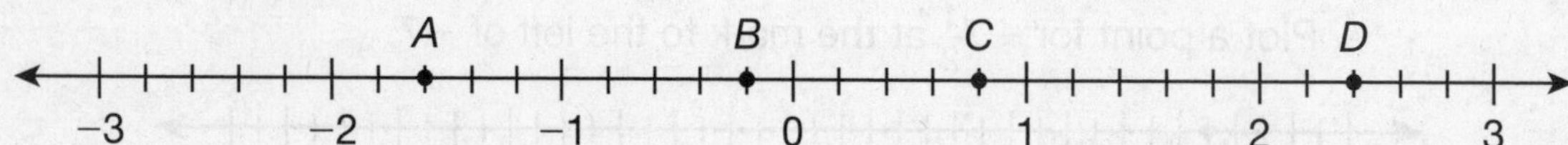

There are ____ spaces between 0 and 1.

The number line is divided into ________. Each mark stands for _____.

$-\frac{1}{5}$ is between _____ and _____. It is located at point ____.

$\frac{4}{5}$ is between _____ and ______. It is located at point ____.

$2\frac{2}{5}$ is between _____ and _____. It is located at point ____.

$-1\frac{3}{5}$ is between _____ and _____. It is located at point ____.

Point ___ represents $-\frac{1}{5}$, point ___ represents $\frac{4}{5}$, point ___ represents $2\frac{2}{5}$, and point ___ represents $-1\frac{3}{5}$.

Lesson Practice

Choose the correct answer.

Use the number line for questions 1 and 2.

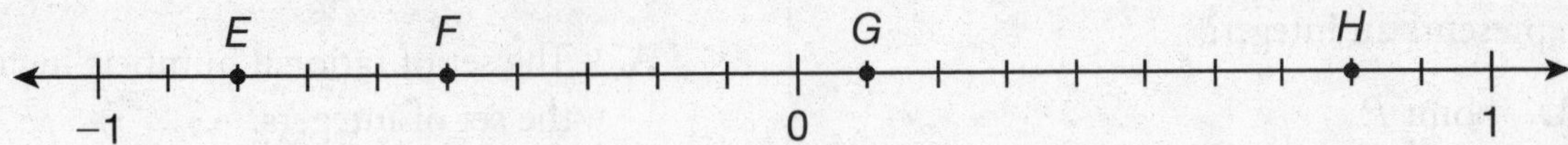

1. Which point on the number line represents −0.5?

   A. point E
   B. point F
   C. point G
   D. point H

2. Which rational number is represented by point E?

   A. $-\frac{8}{10}$
   B. −0.5
   C. $\frac{1}{10}$
   D. 0.8

Use the number line for questions 3 and 4.

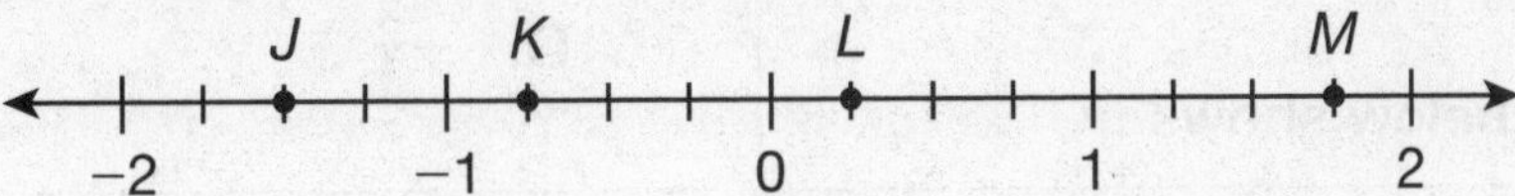

3. Which point on the number line represents $\frac{1}{4}$?

   A. point J
   B. point K
   C. point L
   D. point M

4. Which rational number is represented by point J?

   A. $-1\frac{3}{4}$
   B. $-1\frac{1}{2}$
   C. $-\frac{3}{4}$
   D. $-\frac{1}{2}$

Use the number line for questions 5 and 6.

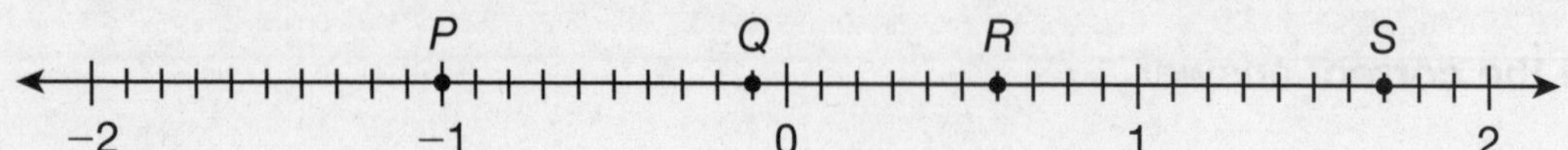

5. Which point on the number line represents an integer?

   A. point *P*

   B. point *Q*

   C. point *R*

   D. point *S*

6. Which rational number is represented by point *R*?

   A. −1.6

   B. −0.6

   C. 0.6

   D. 1.6

7. Which statement is **not** true?

   A. The set of rational numbers includes the set of integers.

   B. All counting numbers are rational numbers.

   C. A mixed number is a rational number.

   D. Zero is not a rational number.

8. What is the opposite of $\frac{3}{8}$?

   A. 38

   B. $\frac{8}{3}$

   C. $-\frac{3}{8}$

   D. $-\frac{8}{3}$

9. Use the sets counting numbers, integers, and rational numbers to answer Part A.

   A. Name the set or sets that each of the numbers below belongs to.

   $-4, \frac{5}{8}, 3,$ and -2.9

   ___

   ___

   ___

   B. Explain why a mixed number is a rational number.

   ___

   ___

   ___

Common Core State Standards: 6.NS.7.a, 6.NS.7.b, 6.NS.7.d

Compare and Order Rational Numbers

Getting the Idea

All rational numbers can be located on a number line. A number line will help you compare and order rational numbers.

To compare numbers, you can use the symbols > **(is greater than)**, < **(is less than)**, or = **(is equal to)**. The expression $p > q$ (*p is greater than q*) means that p is located to the right of q on a number line. The expression $p < q$ (*p is less than q*) means that p is located to the left of q on a number line.

Example 1

Walter and four friends decided to compare the balances in their bank accounts. The table below shows each person's balance.

Bank Balances

Person	Account Balance
Walter	\$35
Ellen	−\$10
Christine	−\$5
Randy	\$40
Peter	−\$20

Order the account balances from greatest to least.

Strategy **Use a number line.**

Step 1 Write each account balance as an integer.

35, −10, −5, 40, −20

Step 2 Plot each integer on a number line.

The negative integers will be to the left of 0. The positive integers will be to the right of 0.

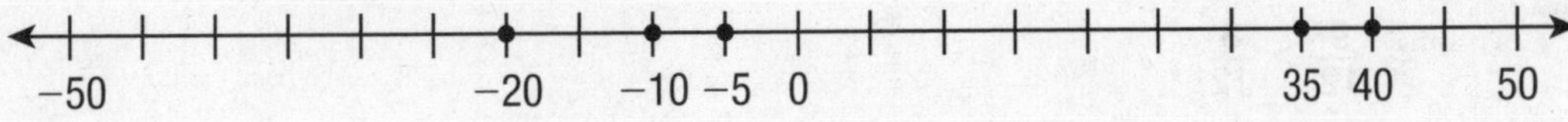

Step 3 Order the integers from greatest to least.

List the integers as they appear on the number line, going from right to left.

40, 35, -5, -10, and -20

Solution **From greatest to least, the account balances are \$40, \$35, −\$5, −\$10, and −\$20.**

In Example 1 you found that -20 is to the left of -5 on a number line, or $-20 < -5$. Now remember that $|-20| = 20$ and $|-5| = 5$. In other words, $|-20|$ is to the right of $|-5|$ on a number line, or $20 > 5$. So in Example 1, Peter's account balance is less than Christine's balance, but his debt is greater than Christine's debt.

To compare and order fractions, you will need fractions with common denominators. One way to find a common denominator is to multiply the denominators of the fractions.

To compare mixed numbers, first look at the whole-number parts. If the whole-number parts are equal, then compare the fraction parts.

Example 2

Which symbol makes this sentence true? Use $>$, $<$, or $=$.

$2\frac{3}{4} \bigcirc 2\frac{2}{3}$

Strategy **Compare the whole-number parts. If necessary, use a common denominator to compare the fraction parts.**

Step 1 Compare the whole-number parts.

$2 = 2$

Step 2 Find a common denominator for the fraction parts.

Multiply the denominators to find a common denominator.

$4 \times 3 = 12$

Step 3 Write the fraction parts as equivalent fractions with a common denominator.

$\frac{3}{4} = \frac{3 \times 3}{4 \times 3} = \frac{9}{12}$

$\frac{2}{3} = \frac{2 \times 4}{3 \times 4} = \frac{8}{12}$

Step 4 Compare the fractions.

$\frac{9}{12} > \frac{8}{12}$

Solution $2\frac{3}{4} \textcircled{>} 2\frac{2}{3}$

When comparing decimals, align the digits on the decimal point, then compare from left to right. The number of decimal places does not affect whether a decimal is greater than or less than another decimal.

Example 3

Kelly owns two Portuguese water dogs. One dog weighs 23.592 kilograms, and the other weighs 23.64 kilograms. Write an expression to compare the dogs' weights. Use $>$, $<$, or $=$.

23.592 ◯ 23.64

Strategy **Align the numbers on the decimal point. Compare from left to right.**

23.592
23.64

Step 1 Compare the tens place.

$2 = 2$, so compare the next greatest place: the ones.

Step 2 Compare the ones place.

$3 = 3$, so compare the next greatest place: the tenths.

Step 3 Compare the tenths.

$5 < 6$

Solution **23.592 ⓒ $<$ 23.64**

Example 4

Last winter, Cedric recorded the low temperature, in degrees Fahrenheit, at his farm over 5 days. His data is shown below.

0.5, 1.3, −2, 1, −1.5

Order the temperatures from lowest to highest.

Strategy **Use a number line.**

Step 1 Plot the numbers on a number line divided into tenths.

Step 2 List the numbers as they appear from left to right on the number line.

−2, −1.5, 0.5, 1, 1.3

This can be written as: $-2 < -1.5 < 0.5 < 1 < 1.3$

Solution **From lowest to highest, the temperatures, in degrees Fahrenheit, are −2, −1.5, 0.5, 1, 1.3**

Coached Example

Order the following numbers from greatest to least:

$-3, 2.6, 3, 2\frac{3}{10}, -3\frac{1}{4}$

Separate the positive numbers from the negative numbers.

The positive numbers are _____, _____, and _____.

Rename 2.6 as a mixed number with a denominator of 10.

2.6 = _____

The greatest positive number is _____.

Compare the remaining two positive numbers. _____ > _____

From greatest to least, the positive numbers are _____, _____, and _____.

The negative numbers are _____ and _____.

Which negative number is greater? _____

From greatest to least, the numbers are _____, _____, _____, _____, and _____.

Lesson Practice

Choose the correct answer.

1. Which list orders the integers from least to greatest?

A. $-5, 3, -2, 4$
B. $-2, 3, 4, -5$
C. $-5, -2, 3, 4$
D. $-2, -5, 3, 4$

2. Which sentence is true?

A. $5\frac{1}{3} < 5\frac{3}{8}$
B. $4\frac{5}{8} > 4\frac{2}{3}$
C. $6\frac{3}{5} = 6\frac{7}{10}$
D. $7\frac{3}{10} > 7\frac{1}{3}$

3. Which list orders the fractions from least to greatest?

A. $\frac{1}{2}, \frac{2}{5}, \frac{3}{4}, \frac{9}{20}$
B. $\frac{3}{4}, \frac{1}{2}, \frac{9}{20}, \frac{2}{5}$
C. $\frac{1}{2}, \frac{3}{4}, \frac{2}{5}, \frac{9}{20}$
D. $\frac{2}{5}, \frac{9}{20}, \frac{1}{2}, \frac{3}{4}$

4. The table shows the distances that four friends live from school.

Distances from School

Student	Distance (in miles)
Teri	$4\frac{3}{4}$
Josie	$4\frac{7}{8}$
Katie	$4\frac{9}{10}$
Ramona	$4\frac{4}{5}$

Which lists the students in order from the greatest distance from school to the least distance?

A. Katie, Ramona, Teri, Josie
B. Katie, Josie, Ramona, Teri
C. Teri, Ramona, Josie, Katie
D. Ramona, Teri, Katie, Josie

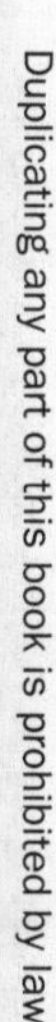

5. The table shows the elevations of various places around the world, in feet below sea level.

Elevations

Place	Elevation (in feet below sea level)
Caspian Sea	92
Dead Sea	1,348
Death Valley	282
Valdes Peninsula	131

Which place has the lowest elevation?

A. Caspian Sea

B. Dead Sea

C. Death Valley

D. Valdes Peninsula

6. Which lists the decimals from greatest to least?

A. 3.917, 39.17, 39.175, 39.7

B. 39.175, 39.17, 3.917, 39.7

C. 3.917, 39.175, 39.17, 39.7

D. 39.7, 39.175, 39.17, 3.917

7. Frank has a bank account balance of −\$45.30. Bruce has a bank account balance of −\$55. Which of the following statements is true?

A. Frank's debt is less than Bruce's.

B. Bruce's debt is less than Frank's.

C. Frank and Bruce have the same debt.

D. Neither Frank nor Bruce have a debt.

8. The table shows the daily high temperatures in Anchorage, Alaska, over a 5-day period in December.

Daily High Temperatures

Day	High Temperature
Monday	−10°F
Tuesday	0°F
Wednesday	−4°F
Thursday	−6°F
Friday	2°F

A. On which days was the temperature the warmest and the coldest?

____

B. Order the temperatures from coldest to warmest.

____

Common Core State Standard:
6.NS.3

Add and Subtract Decimals

Getting the Idea

Adding and subtracting decimals is like adding and subtracting integers. Write the problem vertically, lining up the decimal points. Add or subtract from right to left. Regroup if necessary.

Example 1

Sheila and Leslie hiked 8.76 kilometers in the morning and 4.29 kilometers in the afternoon. How many kilometers did they hike in all?

Strategy **Align the decimal points. Add from right to left. Regroup if necessary.**

Step 1 Align the decimal points.

Place the decimal point in the sum.

$$\begin{array}{r} 8.76 \\ +\ 4.29 \\ \hline . \end{array}$$

Step 2 Add the hundredths.

6 + 9 = 15

Write the 5.

Regroup 10 hundredths as 1 tenth.

$$\begin{array}{r} 1 \\ 8.76 \\ +\ 4.29 \\ \hline .\ 5 \end{array}$$

Step 3 Add the tenths.

1 + 7 + 2 = 10

Write the 0.

Regroup 10 tenths as 1 one.

$$\begin{array}{r} 1\ 1 \\ 8.76 \\ +\ 4.29 \\ \hline .05 \end{array}$$

Step 4 Add the ones.

1 + 8 + 4 = 13

$$\begin{array}{r} 1\ 1 \\ 8.76 \\ +\ 4.29 \\ \hline 13.05 \end{array}$$

Solution **Sheila and Leslie hiked a total of 13.05 kilometers.**

Writing one or more zeros to the right of the last digit in a decimal does not change the value of the decimal. For example:

$1.2 = 1.20 = 1.200$

You can use this to help you add and subtract decimals with different numbers of places.

Example 2

A plumber has two metal pipes. The first pipe is 2.35 meters long. The second pipe is 1.725 meters long. How much longer is the first pipe than the second pipe?

Strategy **Subtract from right to left. Regroup if necessary.**

Step 1 Align the decimal points. Write a 0 in the thousandths place for 2.35.

Place the decimal point in the difference.

$$\begin{array}{r} 2.350 \\ -\,1.725 \\ \hline . \end{array}$$

Step 2 Regroup 1 hundredth as 10 thousandths. Subtract the thousandths.

$$\begin{array}{r} \,4\;10 \\ 2.3\,\not{5}\,\not{0} \\ -\,1.7\;2\;5 \\ \hline .\;5 \end{array}$$

Step 3 Subtract the hundredths.

$$\begin{array}{r} \,4\;10 \\ 2.3\,\not{5}\,\not{0} \\ -\,1.7\;2\;5 \\ \hline .\;2\;5 \end{array}$$

Step 4 Regroup 1 one as 10 tenths. Subtract the tenths.

$$\begin{array}{r} 1\;13\;4\;10 \\ \not{2}.\not{3}\,\not{5}\,\not{0} \\ -\,1.7\;2\;5 \\ \hline .6\;2\;5 \end{array}$$

Step 5 Subtract the ones.

$$\begin{array}{r} 1\;13\;4\;10 \\ \not{2}.\not{3}\,\not{5}\,\not{0} \\ -\,1.7\;2\;5 \\ \hline 0.6\;2\;5 \end{array}$$

Solution **The first pipe is 0.625 meter longer than the second pipe.**

Example 3

Liana and Terrence each bought orange juice for a school brunch. Liana bought a container with 1.89 liters of orange juice. Terrence bought 3 small containers, each of which held 0.473 liter of orange juice. Who bought more orange juice? How much more orange juice, in liters, did that student buy?

Strategy **Add. Compare. Then subtract.**

Step 1 Add to find the total number of liters Terrence bought.

$$\begin{array}{r} {\scriptstyle 1\,2} \\ 0.473 \\ 0.473 \\ +\ 0.473 \\ \hline 1.419 \end{array}$$

Step 2 Compare Liana's total to Terrence's total.

Liana bought 1.89 liters. Terrence bought 1.419 liters.

1.89 > 1.419, so Liana bought more orange juice than Terrence.

Step 3 Subtract Terrence's total from Liana's total.

First write a 0 in the thousandths place for 1.89.

$$\begin{array}{r} {\scriptstyle 8\,10} \\ 1.8\not{9}\not{0} \\ -\ 1.419 \\ \hline 0.471 \end{array}$$

Solution **Liana bought 0.471 liter of orange juice more than Terrence did.**

Coached Example

Melanie had a budget of $225 for costumes for the school play. She spent $86.56 on jackets and $45.38 on hats. How much money does Melanie have left in the budget?

First add to find how much money Melanie has spent so far.

The jackets cost $__________.

The hats cost $__________.

Add. Show your work in the space below.

Then subtract the total amount spent from $225.

Remember that 225 = 225.00

Subtract. Show your work in the space below.

Melanie has $____________ left in the budget.

Lesson Practice

Choose the correct answer.

1. Debra bought two bottles of conditioner at the hair salon. One bottle contained 0.355 liter (L) and the other contained 0.877 L. How many liters of conditioner did she buy in all?

A. 1.122 L
B. 1.232 L
C. 9.225 L
D. 12.32 L

2. Mr. Farmer has a greyhound that can run 37.35 miles per hour. He also has a quarter horse that can run 47.5 miles per hour. How much faster can the quarter horse run than the greyhound?

A. 9.2 miles per hour
B. 10.15 miles per hour
C. 11.45 miles per hour
D. 11.85 miles per hour

3. Mr. Palmer had $5,675.68 in his savings account. He then deposited $2,168.79 more in his account. How much is in his savings account now?

A. $7,844.47
B. $7,843.37
C. $7,734.47
D. $7,733.37

4. A marathon is a race with a distance of 26.2 miles (mi). Lauren is competing in a marathon and has run 10.75 mi so far. How many more miles does she need to run to complete the marathon?

A. 8.13 mi
B. 15.27 mi
C. 15.45 mi
D. 15.55 mi

5. The mean distance from Mars to the sun is 141.633 million miles. The mean distance from Mercury to the sun is 35.983 million miles. Approximately how much closer to the sun is Mercury than Mars?

A. 218.197 million miles
B. 177.616 million miles
C. 142.35 million miles
D. 105.65 million miles

6. In the 2010 Winter Olympics 500-meter mens' speed skating finals, the gold medalist's best time was 40.77 seconds and the silver medalist's best time was 40.821 seconds. How many seconds faster was the gold medalist's best time than the silver medalist's best time?

A. 0.051 second
B. 0.114 second
C. 0.744 second
D. 36.744 seconds

7. A blue piece of string is 2.355 meters long. A red piece of string is 3.8 meters long. How much longer is the red piece of string than the blue piece of string?

A. 1.975 meters
B. 1.445 meters
C. 0.725 meter
D. 0.653 meter

8. Claire's family drove 129.5 miles on the first day of their vacation. The next day they drove 43.25 miles, and the third day they drove 36.5 miles. How many more miles did they drive on the first day than on the second and third days combined?

A. 211.25 miles
B. 81.75 miles
C. 49.75 miles
D. 43.00 miles

9. The table below shows the masses of three U.S. coins.

Masses of U.S. Coins

Coin	Mass (in grams)
Dime	2.268
Quarter	5.67
Half dollar	11.34

A. A dime and a quarter are placed on one side of a balance scale. A half dollar is placed on the other side. Which side of the scale has the greater mass, in grams? How much greater is it? Show or explain how you determined your answer.

B. What is the total mass of all three coins? Show or explain how you determined your answer.

Common Core State Standard:
6.NS.3

Multiply and Divide Decimals

Getting the Idea

Multiplying decimals is similar to whole numbers. Multiply the factors as if there were no decimal points. Then, add the number of decimal places in the factors. That sum is the number of decimal places in the product.

Example 1

What is the product of 0.4 × 0.6?

Strategy **Multiply the factors. Then place the decimal point in the product.**

Step 1 Write the problem vertically.

$$\begin{array}{r} 0.4 \\ \times\ 0.6 \\ \hline \end{array}$$

Step 2 Multiply as you would with whole numbers.

$$\begin{array}{r} {\scriptstyle 2} \\ 0.4 \\ \times\ 0.6 \\ \hline 24 \end{array}$$

You do not need to multiply by 0 since the product would be 0.

Step 3 Add the number of decimal places in the factors. Insert the decimal point in the product.

$$\begin{array}{rl} {\scriptstyle 2} & \\ 0.4 & \leftarrow \text{1 decimal place} \\ \times\ 0.6 & \leftarrow \text{1 decimal place} \\ \hline 0.24 & \leftarrow \text{1 + 1, or 2, decimal places} \end{array}$$

Solution **0.4 × 0.6 = 0.24**

Example 2

Karin buys fancy yarn at a cost of \$0.36 per yard. She uses 0.5 yard of that yarn to make a necklace. How much did the yarn used to make the necklace cost?

Strategy **Multiply the factors. Then place the decimal point in the product.**

```
     3
  0.36   ← 2 decimal places
× 0.5    ← 1 decimal place
 0.180   ← 3 decimal places
```

Drop the zero after the 8 since this is an amount of money:

0.180 = 0.18

Solution **The cost of the yarn used to make the necklace is \$0.18.**

Example 3

Multiply.

```
 1.748
× 4.6
```

Strategy **Multiply the factors. Then place the decimal point in the product.**

Step 1 Multiply by the 6 tenths.

```
  4 24
 1.748
× 4.6
10488
```

Step 2 Multiply by the 4 ones.

Write a 0 in the partial product so that you begin writing the partial product in the correct place. Then multiply.

```
  2 13
  4 24
 1.748
× 4.6
10488
69920
```

Step 3 Add the partial products and insert the decimal point.

```
    213
    424
  1.748      ← 3 decimal places
 × 4.6       ← 1 decimal place
  10488
+ 69920
 8.0408      ← 4 decimal places
```

Solution **1.748 × 4.6 = 8.0408**

Dividing decimals is also similar to dividing integers. One important difference is that division with decimals has no remainder.

When dividing a decimal by an integer, start by placing the decimal point in the quotient. Then divide as you would with integers. If necessary, insert zeros after the last nonzero digit in the dividend until you find a decimal quotient that terminates or repeats.

Example 4

A scientist has 18.6 milliliters of a chemical solution. She divides the solution evenly among 8 different cylinders for an experiment. How many milliliters of solution will be in each cylinder?

Strategy **Write the problem vertically. Divide from left to right. If necessary, insert zeros in the dividend.**

Step 1 Write the problem vertically. Place the decimal point in the quotient.

```
     .
 8)18.6
```

Step 2 Divide as you would divide integers.

```
    2.3
 8)18.6
  −16        ← Multiply: 2 × 8 = 16
    2 6      ← Subtract: 18 − 16 = 2. Bring down the 6.
   −2 4      ← Multiply: 3 × 8 = 24
      2      ← Subtract: 26 − 24 = 2
```

Step 3 There is a remainder of 2. Insert a 0 in the dividend and continue dividing.

```
    2.32
 8)18.60
  -16
    2 6
   -2 4
      20   ← Bring down the 0.
     -16   ← Multiply: 2 × 8 = 16
       4   ← Subtract: 20 − 16 = 4
```

Step 4 There is a remainder of 4. Insert another 0 in the dividend and continue dividing.

```
    2.325
 8)18.600
  -16
    2 6
   -2 4
      20
     -16
       40   ← Bring down the 0.
      -40   ← Multiply: 5 × 8 = 40
        0   ← Subtract: 40 − 40 = 0
```

There is no remainder. The quotient is a terminating decimal.

Solution **Each cylinder will have 2.325 milliliters of the chemical solution.**

To divide by a decimal, change the decimal to an integer by moving the decimal point to the right. Then move the decimal point in the dividend the same number of places to the right.

For example, to divide $6.825 \div 1.75$, move the decimal point two places to the right in both numbers. This is the same as multiplying both numbers by 100.

$1.75 \times 100 = 175$

$6.825 \times 100 = 682.5$

The quotient of $682.5 \div 175$ is the same as the quotient of $682.5 \div 1.75$.

Example 5

Divide: 94.575 ÷ 3.9

Strategy **Change the divisor to an integer. Move the decimal point in the dividend the same number of places. Then divide from left to right.**

Step 1 Change the divisor to an integer.

3.9 → 39

The decimal point moved 1 place to the right. This is the same as multiplying the divisor by 10.

Step 2 Move the decimal point in the dividend 1 place to the right. This is the same as multiplying the dividend by 10.

94.575 → 945.75

Step 3 Rewrite the problem with the new dividend and divisor.

Place the decimal point in the quotient.

```
      .
39)945.75
```

Step 4 Divide each place from left to right.

```
     24.25
39)945.75
  -78
   165
  -156
     97
    -78
     195
    -195
       0
```

Solution **94.575 ÷ 3.9 = 24.25**

Coached Example

Mr. Giamatti bought 7.5 pounds of tea leaves. He spent a total of $142.50. How much does each pound of tea leaves cost?

Change the divisor, 7.5, to a whole number.

Move the decimal point _____ place(s) to the right. Multiply by _________.

7.5 × _____ = _____

Move the decimal point in the dividend the same number of places to the right.

142.5 × _____ = _____

Write the problem vertically with the new divisor and dividend. Then divide. Show your work.

The quotient is ___________.

Each pound of tea leaves costs $____________.

Lesson Practice

Choose the correct answer.

1. Multiply: 0.7×0.4

   A. 0.28
   B. 2.8
   C. 28
   D. 280

2. Divide: $50.28 \div 12$

   A. 4.19
   B. 4.21
   C. 41.9
   D. 42.1

3. Maxim raised $890.88 for charity. He divided the amount equally among his sixteen favorite charities. How much did each charity receive?

   A. $41.61
   B. $54.16
   C. $55.18
   D. $55.68

4. Kristina rides her bicycle 13.25 miles to and from her job each week. How many miles does she bike in all to and from her job in 29 weeks?

   A. 3.8425 miles
   B. 38.425 miles
   C. 384.25 miles
   D. 3,842.5 miles

5. What is the product of 3.456×1.3?

   A. 0.4928
   B. 4.4928
   C. 44.928
   D. 449.28

6. What is the quotient when 2.375 is divided by 0.05?

   A. 4.75
   B. 5.55
   C. 44.35
   D. 47.5

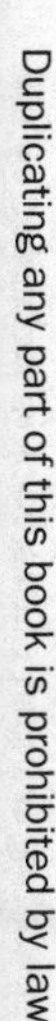

7. The cost of Ruben's large storage unit is $285.47 per month. If Ruben pays in advance for 2 years of storage, how much will he pay in all?

A. $1,712.82
B. $6,851.28
C. $6,951.28
D. $7,851.28

8. A road race is 10 kilometers long. After the starting line, there are water stations every 0.4 kilometer, including at the finish line. How many water stations are there?

A. 25
B. 40
C. 250
D. 400

9. Greg's Gas Station sells three different kinds of gasoline: regular, plus, and premium.

A. Mr. Adams spent $36.25 on 12.5 gallons of regular gasoline at Greg's Gas Station. Determine the cost per gallon for regular gasoline, showing each step in the process.

B. At Greg's Gas Station, premium gasoline costs $0.14 more per gallon than regular gasoline. How much would Mr. Adams have paid if he bought 12.5 gallons of premium gasoline instead? Show or explain your work.

Common Core State Standard:
6.NS.1

Divide Fractions and Mixed Numbers

Getting the Idea

To divide a number by a fraction, multiply the number by the **reciprocal** of the fraction. Two numbers are reciprocals if their product is 1.

To find the reciprocal of a fraction, switch its numerator and denominator. For example, the reciprocal of $\frac{3}{5}$ is $\frac{5}{3}$ since $\frac{3}{5} \times \frac{5}{3} = 1$.

You can use models to help you divide fractions.

Example 1

Jeffrey has a wooden board that is $\frac{2}{3}$ yard long. He wants to cut the board into pieces that are $\frac{1}{9}$ yard long. How many pieces will Jeffrey cut?

Strategy **Write a division sentence. Use models to find the quotient.**

Step 1 Write a division sentence to represent the situation.

Let p represent the number of $\frac{1}{9}$-yard pieces Jeffrey will cut.

$\frac{2}{3} \div \frac{1}{9} = p$

Step 2 Draw a rectangle. Divide it into 3 equal parts. Shade the model to represent $\frac{2}{3}$.

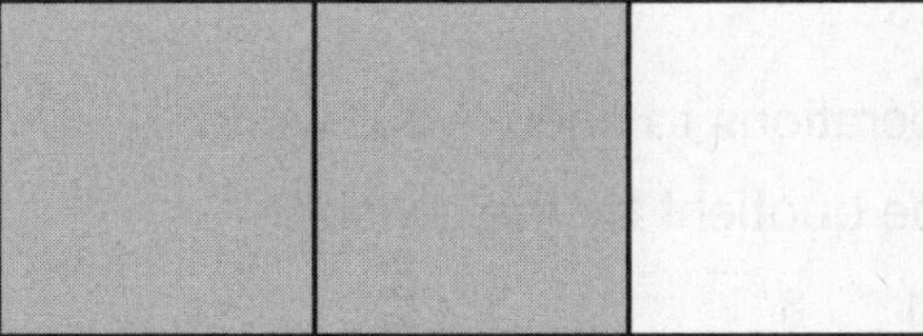

Step 3 Find the number of $\frac{1}{9}$s in $\frac{2}{3}$.

Divide the rectangle into 9 equal parts.

Step 4 Count the number of shaded sections.

There are 6 shaded sections.

Solution **Jeffrey will cut the board into 6 pieces.**

Remember that multiplication and division are **inverse operations**. For example, the division sentence $\frac{3}{4} \div \frac{1}{12} = 9$ is related to the multiplication sentence $\frac{1}{12} \times 9 = \frac{3}{4}$. You can use related multiplication sentences to check your work in a division sentence.

Example 2

$\frac{3}{10} \div \frac{5}{6} = \square$

Strategy **Find the reciprocal of the divisor and multiply.**

Step 1 Write the reciprocal of the divisor, $\frac{5}{6}$.

$\frac{5}{6} \times \frac{6}{5} = 1$, so the reciprocal of $\frac{5}{6}$ is $\frac{6}{5}$.

Step 2 Rewrite the division as multiplication.

$\frac{3}{10} \div \frac{5}{6} = \frac{3}{10} \times \frac{6}{5}$

Step 3 Simplify and multiply.

Simplify: divide 10 and 6 by the greatest common factor (GCF), 2.
Multiply.

$$\frac{3}{10} \times \frac{6}{5} = \frac{3}{\underset{5}{\cancel{10}}} \times \frac{\overset{3}{\cancel{6}}}{5} = \frac{9}{25}$$

Step 4 Use inverse operations to check your work.

Multiply the quotient by the divisor.

$$\frac{5}{6} \times \frac{9}{25} = \frac{\overset{1}{\cancel{5}}}{\underset{2}{\cancel{6}}} \times \frac{\overset{3}{\cancel{9}}}{\underset{5}{\cancel{25}}} = \frac{3}{10} \checkmark$$

Solution $\mathbf{\frac{3}{10} \div \frac{5}{6} = \frac{9}{25}}$

Example 3

Marrell brought back $\frac{3}{4}$ pound of chocolate from her vacation in Belgium. She gave $\frac{1}{8}$ pound of chocolate to each of her nephews and nieces. If Marrell gave away all the chocolate, how many nephews and nieces does she have?

Strategy **Write a division sentence. Find the reciprocal of the divisor and multiply.**

Step 1 Write a division sentence to represent the situation.

Let n represent the number of nephews and nieces.

$\frac{3}{4} \div \frac{1}{8} = n$

Step 2 Write the reciprocal of the divisor, $\frac{1}{8}$.

$\frac{1}{8} \times 8 = 1$, so the reciprocal of $\frac{1}{8}$ is 8.

Step 3 Rewrite the division as multiplication.

$\frac{3}{4} \div \frac{1}{8} = \frac{3}{4} \times 8$

Step 4 Simplify and multiply.

Divide 4 and 8 by the greatest common factor (GCF), 4.

$\frac{3}{4} \times 8 = \frac{3}{\cancel{4}_1} \times \frac{\cancel{8}^2}{1} = 6$

Step 5 Use inverse operations to check your work.

Multiply the quotient by the divisor.

$\frac{1}{8} \times 6 = \frac{1}{\cancel{8}_4} \times \frac{\cancel{6}^3}{1} = \frac{3}{4}$ ✓

Solution **Marrell has 6 nephews and nieces.**

Another way to solve a problem like Example 3 is to rewrite the division problem as one fraction and then simplify:

$$\frac{(\text{numerator of the dividend} \times \text{denominator of the divisor})}{(\text{denominator of the dividend} \times \text{numerator of the divisor})}.$$

In Example 3:

$\frac{3}{4} \div \frac{1}{8} = \frac{3 \times 8}{4 \times 1} = \frac{24}{4} = 6$

This is the same answer you found above.

In general, for division of fractions:

$\frac{a}{b} \div \frac{c}{d} = \frac{ad}{bc}$

To divide mixed numbers, rename the mixed numbers as improper fractions, and then multiply the dividend by the reciprocal of the divisor.

Example 4

Jessica worked a total of $8\frac{1}{2}$ hours at the local soup kitchen as a volunteer over the holidays. She worked $2\frac{3}{4}$ hours on her first day. What fraction of the total time did Jessica work that first day?

Strategy **Write a division sentence. Rename mixed numbers as improper fractions. Then solve.**

Step 1 Write a division sentence to represent the situation.

Let *f* represent the fraction of the total time she worked on the first day.

$$2\frac{3}{4} \div 8\frac{1}{2} = f$$

Step 2 Rename the mixed numbers as improper fractions.

$$2\frac{3}{4} = \frac{11}{4} \text{ and } 8\frac{1}{2} = \frac{17}{2}$$

$$2\frac{3}{4} \div 8\frac{1}{2} = \frac{11}{4} \div \frac{17}{2}$$

Step 3 Rewrite the division problem as a multiplication problem, using the reciprocal of the divisor.

$$\frac{11}{4} \div \frac{17}{2} = \frac{11}{4} \times \frac{2}{17}$$

Step 4 Simplify and multiply.

Divide 4 and 2 by the greatest common factor (GCF), 2.

$$\frac{11}{4} \times \frac{2}{17} = \frac{11}{\underset{2}{\cancel{4}}} \times \frac{\overset{1}{\cancel{2}}}{17} = \frac{11}{34}$$

Step 5 Use inverse operations to check your work.

Multiply the quotient by the divisor.

$$\frac{17}{2} \times \frac{11}{34} = \frac{\overset{1}{\cancel{17}}}{2} \times \frac{11}{\underset{2}{\cancel{34}}} = \frac{11}{4} = 2\frac{3}{4} \checkmark$$

Solution **Jessica worked $\frac{11}{34}$ of the total number of hours she worked at the soup kitchen on her first day.**

Coached Example

Mrs. Castillo's son and his friends were helping her clean up the backyard after a storm. She made $\frac{7}{8}$ gallon of lemonade for her son and his friends. If each serving of lemonade is 1 cup, how many servings can Mrs. Castillo offer? (Note: 1 cup = $\frac{1}{16}$ gallon)

Write a division sentence to represent the situation.
Let *s* represent the number of servings.

$\_\_\_\_ \div \_\_\_\_ = s$

Write the reciprocal of the divisor.

$\frac{1}{16} \times \_\_\_\_ = 1$, so the reciprocal of $\frac{1}{16}$ is ____.

Rewrite the division as multiplication.

$\frac{7}{8} \times \_\_\_\_$

Multiply.

$\frac{7}{8} \times \frac{16}{1} = \frac{7 \times \_\_\_}{\_\_\_ \times 1} = \_\_\_\_ = \_\_\_\_$

Use inverse operations to check your work.

Multiply the quotient by the divisor.

$\_\_\_\_ \times \frac{1}{16} = \frac{\_\_\_ \times 1}{1 \times \_\_\_} = \_\_\_\_ = \_\_\_\_$

Mrs. Castillo can offer ____ servings of lemonade.

Lesson Practice

Choose the correct answer.

1. Hassan drew the model below to represent a division sentence.

Which of the following division sentences does the model represent?

A. $\frac{1}{2} \div \frac{1}{6}$

B. $\frac{1}{4} \div \frac{2}{3}$

C. $\frac{3}{4} \div \frac{2}{3}$

D. $\frac{3}{4} \div \frac{1}{12}$

2. $\frac{1}{20} \div \frac{4}{5} = \square$

A. $\frac{1}{25}$

B. $\frac{1}{20}$

C. $\frac{1}{16}$

D. 25

3. $10\frac{1}{2} \div 3\frac{1}{5} = \square$

A. $1\frac{23}{32}$

B. $3\frac{9}{32}$

C. $3\frac{3}{5}$

D. $5\frac{1}{2}$

4. What is the reciprocal of 4?

A. -4

B. 0

C. $\frac{1}{4}$

D. $|4|$

5. What is the reciprocal of $4\frac{5}{8}$?

A. $\frac{8}{45}$

B. $\frac{8}{37}$

C. $\frac{8}{5}$

D. $3\frac{7}{8}$

6. Which shows how you can check that $\frac{5}{8} \div \frac{2}{3} = \frac{15}{16}$?

A. $\frac{15}{16} \div \frac{2}{3} = \frac{5}{8}$

B. $\frac{15}{16} \div \frac{5}{8} = \frac{2}{3}$

C. $\frac{3}{2} \times \frac{15}{16} = \frac{5}{8}$

D. $\frac{2}{3} \times \frac{15}{16} = \frac{5}{8}$

7. Joe is making a recipe that calls for $\frac{3}{4}$ teaspoon of cinnamon. His only measuring spoon holds $\frac{1}{8}$ teaspoon. How many times will he need to fill his measuring spoon to get enough cinnamon for the recipe?

A. $\frac{3}{32}$

B. 3

C. 6

D. 12

8. Diego practices guitar for a total of $9\frac{3}{4}$ hours each week. He practices for $\frac{3}{4}$ hour each time. How many times does Diego practice guitar each week?

A. 13

B. 9

C. 6

D. 3

9. Vera estimates that it will take her $16\frac{2}{3}$ hours to complete a project for her playwriting class. She spent $4\frac{1}{6}$ hours working on the project last weekend. What fraction of the time needed to complete the project did she work last weekend?

A. Solve the problem. Show your work.

B. Explain how to check that the quotient you got in Part A is correct.

___

___

___

Common Core State Standards: 6.NS.6.b, 6.NS.6.c

The Coordinate Plane

Getting the Idea

You can use a **coordinate plane** to locate points. A coordinate plane is formed by a horizontal number line, called the ***x*-axis**, and a vertical number line, called the ***y*-axis**. Each axis includes both positive and negative numbers. The coordinate plane is divided into four sections called **quadrants**. They are numbered with Roman numerals in a counterclockwise direction, as shown below.

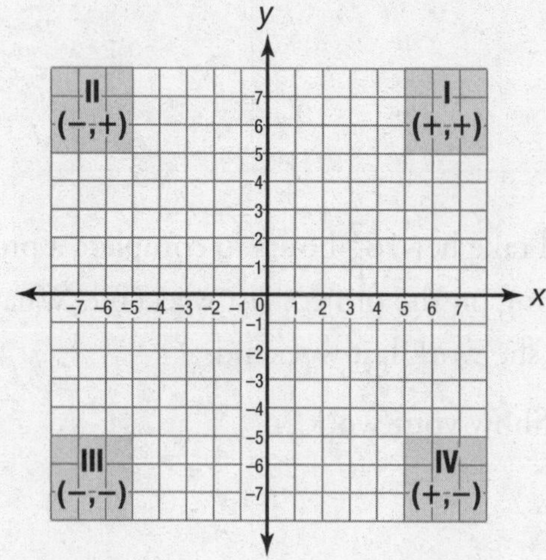

An **ordered pair** of numbers in the form (x, y) names a point on a coordinate plane. The first number of the ordered pair is the ***x*-coordinate**. It tells how many units to move to the left or the right of the **origin**, point (0, 0). The second number is the ***y*-coordinate**. It tells how many units to move up or down from the origin.

By looking at whether the *x*- and *y*-coordinates are positive or negative, you can tell which quadrant contains a given point without seeing it graphed on a coordinate plane. Use the table below to help you.

Quadrant	*x*-coordinate	*y*-coordinate
I	+	+
II	−	+
III	−	−
IV	+	−

Points on the *x*-axis or the *y*-axis are not in any quadrant.

Example 1

Plot $(-4, 6)$ on the coordinate plane. Label the point A.

Strategy **Use ordered pairs to plot a point.**

Step 1 Use the signs of the coordinates to find the quadrant for point A.

The coordinates for point A are (negative, positive), or $(-, +)$.
Point A will be in quadrant II.

Step 2 Start at the origin. Find the x-coordinate for point A.

The x-coordinate is -4.
Move 4 units to the left.

Step 3 From -4 on the x-axis, find the y-coordinate for point A.

The y-coordinate is 6.
Move up 6 units and label point A.

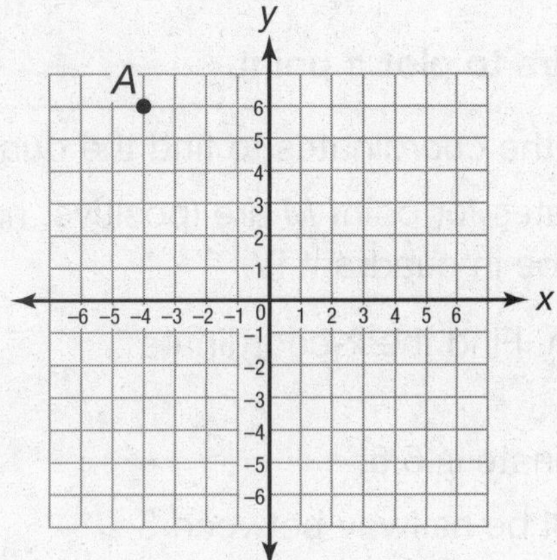

Solution **Point A is shown on the coordinate plane above.**

Example 2

Jody used a coordinate grid to map where she planted each type of vegetable. What ordered pair tells where Jody planted lettuce?

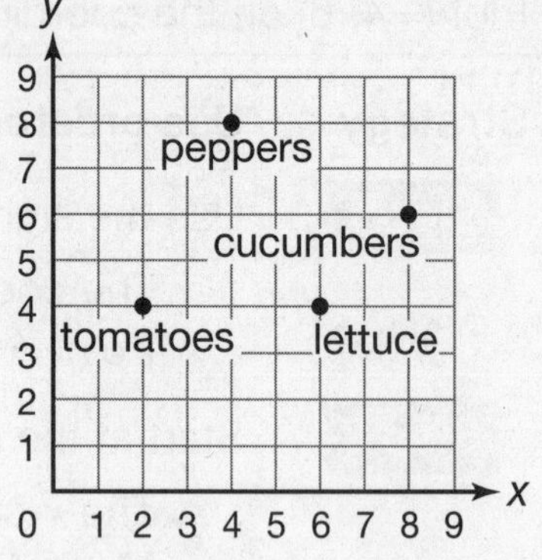

Strategy **Locate the point on the plane. Find the coordinates.**

Lettuce lines up with 6 on the *x*-axis and 4 on the *y*-axis.

Its coordinates are (6, 4).

Solution **Jody planted lettuce at (6, 4).**

Example 3

Plot (5.5, −3.5) on a coordinate plane. Label the point *M*.

Strategy **Use ordered pairs to plot a point.**

Step 1 Use the signs of the coordinates to find the quadrant for point *M*.

The coordinates for point *M* are (positive, negative), or (+, −). Point *M* will be in quadrant IV.

Step 2 Start at the origin. Find the *x*-coordinate for point *M*.

The *x*-coordinate is 5.5.

The point will be halfway between 5 and 6 on the *x*-axis.

Step 3 From 5.5 on the *x*-axis, find the *y*-coordinate for point *M*.

The *y*-coordinate is −3.5.

The point will be halfway between −3 and −4 on the *y*-axis.

Notice that point *M* is **not** on any of the grid lines of the coordinate plane.

Solution **Point *M* is shown on the coordinate plane above.**

Coached Example

The coordinate plane below represents the streets in Brad and Cara's town.

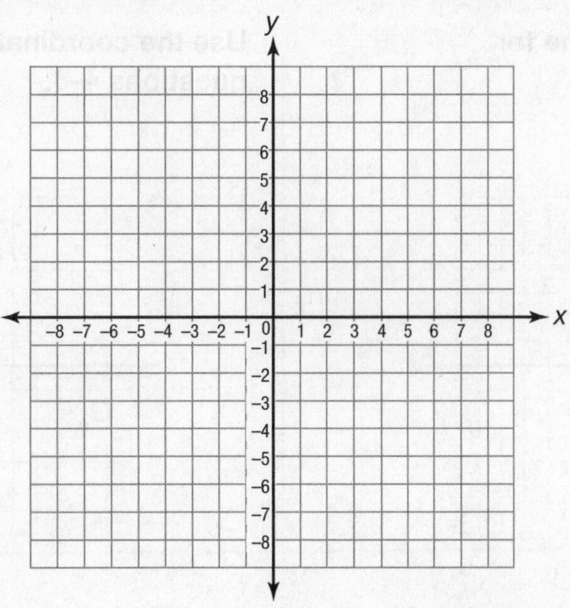

Brad's house is at $(-6, 6\frac{1}{2})$ and Cara's house is at $(4\frac{1}{2}, -5)$. Plot and label the points for both houses.

Start with Brad's house.

The ____-coordinate is negative and the ____-coordinate is positive.

The point for Brad's house will be in quadrant ______.

Start at the origin, which is the point (_____, _____).

Move 6 units to the ___________ of the origin.

From _____ on the *x*-axis, move __________ $6\frac{1}{2}$ units.

Plot the point and label it "*B*" for Brad.

Now locate Cara's house.

The ____-coordinate is positive and the ____-coordinate is negative.

The point for Cara's house will be in quadrant ______.

Start at the origin and move $4\frac{1}{2}$ units to the ___________.

From _____ on the *x*-axis, move __________ 5 units.

Plot the point and label it "*C*" for Cara.

Lesson Practice

Choose the correct answer.

Use the coordinate plane for questions 1–3.

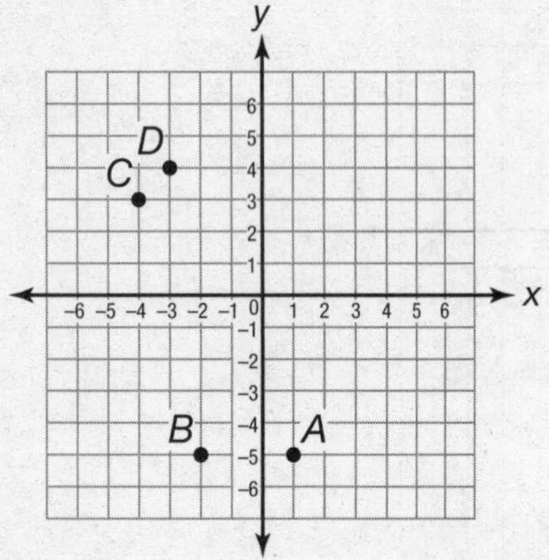

1. Which point is located at $(-2, -5)$?
   A. point A
   B. point B
   C. point C
   D. point D

2. Which point is located at $(-4, 3)$?
   A. point A
   B. point B
   C. point C
   D. point D

3. Which point is located in quadrant IV?
   A. point A
   B. point B
   C. point C
   D. point D

Use the coordinate plane for questions 4–6.

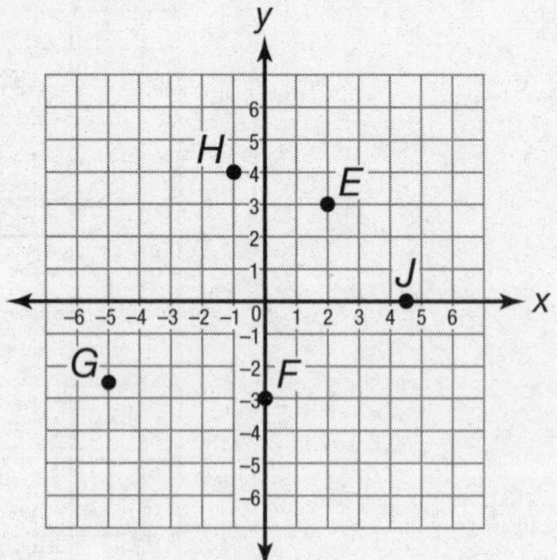

4. Which ordered pair names the location of point J?
   A. $(0, -4.5)$
   B. $(0, 4.5)$
   C. $(-4.5, 0)$
   D. $(4.5, 0)$

5. Which point is located at $(-5, -2\frac{1}{2})$?
   A. point E
   B. point F
   C. point G
   D. point H

6. In which quadrant is point H located?
   A. quadrant I
   B. quadrant II
   C. quadrant III
   D. quadrant IV

7. The x- and y-coordinates of point N are both negative. In which quadrant is point N located?

A. quadrant I

B. quadrant II

C. quadrant III

D. quadrant IV

8. Point V is located at (5.2, −7.3). In which quadrant is point V located?

A. quadrant I

B. quadrant II

C. quadrant III

D. quadrant IV

9. Use the coordinate plane below.

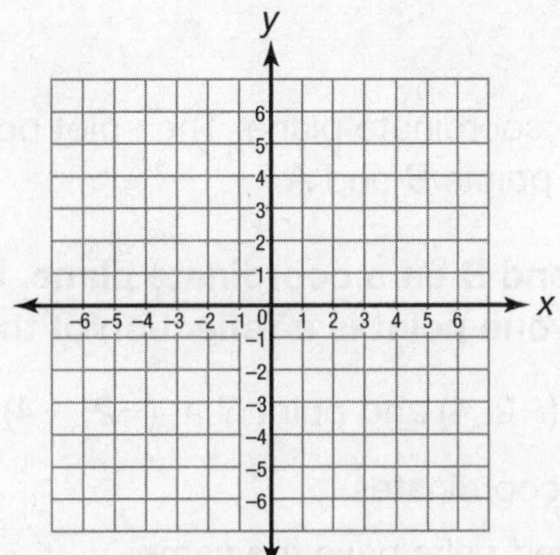

A. Plot and label point P at (3, −4).

B. Plot a point in quadrant II. Label it point B. What are the coordinates of point B?

Common Core State Standards: 6.NS.6.b, 6.NS.8

Solve Problems in the Coordinate Plane

Getting the Idea

A point can be flipped over a line. The new point is called the **reflection** of the original point. When a point is reflected across the *x*-axis, the *y*-axis, or both axes on a coordinate plane, the signs of one or both coordinates will change. The new point also will be in a different quadrant.

Example 1

Plot point *A* at (−2, 4) on the coordinate plane. Then plot point *B* at (−2, −4). Compare the coordinates and locations of points *B* and *A*.

Strategy **Plot points *A* and *B* on a coordinate plane. Compare their coordinates. Describe how one point is a reflection of the other.**

Step 1 Plot point *A* at (−2, 4) and point *B* at (−2, −4).

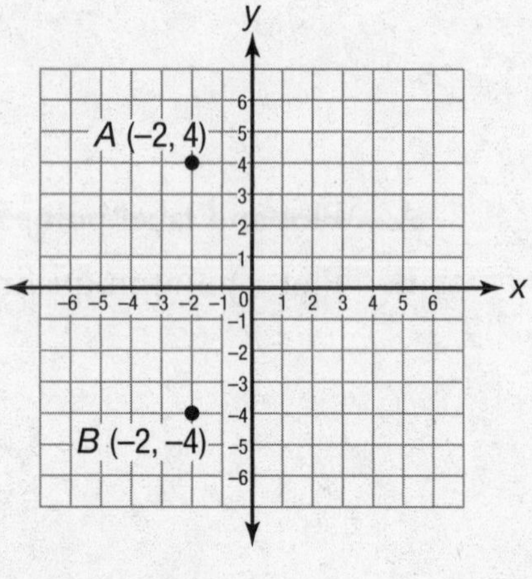

Step 2 Compare their coordinates.

Both ordered pairs have the same *x*-coordinate, −2.

The *y*-coordinates of the ordered pairs are 4 and −4.

They have different signs.

Step 3 Describe the locations of the points.

Both points lie on the same vertical line, $x = -2$.

Point *A* is in quadrant II and is 4 units *above* the *x*-axis.

Point *B* is in quadrant III and is 4 units *below* the *x*-axis.

So, point *B* is a reflection of point *A* over the *x*-axis.

If you folded the grid along the *x*-axis, point *A* would fold onto point *B*.

Solution **The coordinate plane above shows points *A* and *B*. The points are reflections of each other across the *x*-axis, and their *y*-coordinates have different signs.**

The statements below show the relationship between reflected points and the signs of their coordinates.

- If a point (x, y) is reflected across the x-axis, the sign of its y-coordinate changes.

$$(x, y) \rightarrow (x, -y)$$

- If a point (x, y) is reflected across the y-axis, the sign of its x-coordinate changes.

$$(x, y) \rightarrow (-x, y)$$

- If a point (x, y) is reflected across both axes, the signs of both of its coordinates change.

$$(x, y) \rightarrow (-x, -y)$$

You can use absolute value to find the distance between two points on a coordinate plane. Each axis on a coordinate plane is a number line. Since the absolute value of a number is its distance from zero on a number line, you can use absolute values to find horizontal and vertical distances on the coordinate plane.

Example 2

A map of some important places in Carr County is shown at the right. What is the distance, in kilometers, between City Hall and the library?

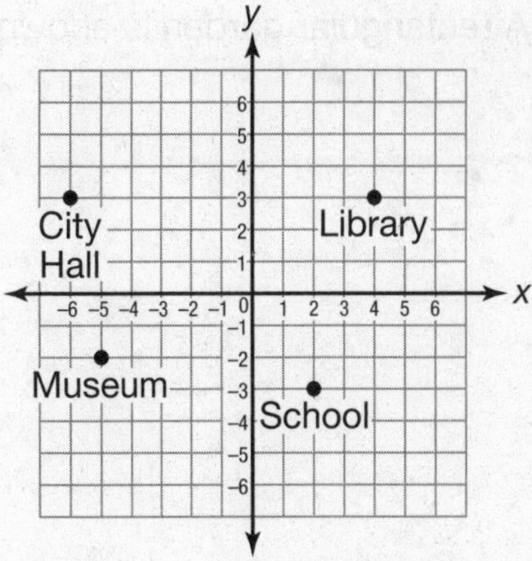

Strategy **Use absolute values to find the distance between points. Then use the scale to find the distance in kilometers.**

Step 1 Identify the ordered pairs.

City Hall is at $(-6, 3)$.
The library is at $(4, 3)$.

Step 2 Use absolute value to find the distance from City Hall to the y-axis.

Think of the line $y = 3$ as a horizontal number line.

Zero on that number line is the point at which $y = 3$ crosses the y-axis, or where $x = 0$.

The x-coordinate for City Hall is -6.

$|-6| = 6$

The distance between City Hall and the y-axis is 6 units.

Step 3 Use absolute value to find the distance from the public library to the y-axis.

The x-coordinate for the library is 4.

$|4| = 4$

The distance between the library and the y-axis is 4 units.

(–6, 3) City Hall | 6 units | 4 units | (4, 3) Library

Scale: ⊢⊣ = 1 kilometer

Step 4 Add the absolute values to find the total distance.

The total distance between City Hall and the library on the map is:

$6 + 4 = 10$ units

Each unit on the map represents 1 kilometer, so the actual distance is 10 kilometers.

Solution **The distance between City Hall and the library is 10 kilometers.**

You can also use absolute values to find the **perimeter** of a figure on a coordinate plane.

Example 3

A rectangular garden is shown below. What is the perimeter of the garden?

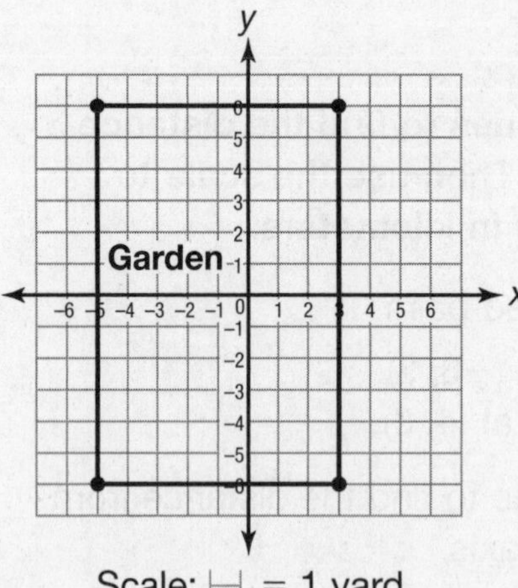

Scale: ⊢⊣ = 1 yard

Strategy **Use absolute values to find the length of one vertical side and one horizontal side. Then find the perimeter.**

Step 1 Use absolute value to find the length of a vertical side.

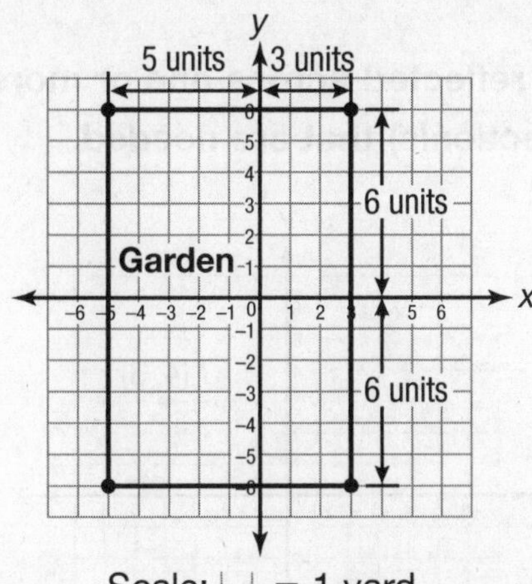

Scale: ⊢⊣ = 1 yard

The endpoints of one vertical side are (3, 6) and (3, −6).

Think of the line $x = 3$ as a vertical number line.

Zero on that number line is the point at which $x = 3$ crosses the x-axis, or (3, 0).

The y-coordinates for the endpoints are 6 and −6.

$|6| = 6$ $|-6| = 6$

The total length of a vertical side is: 6 + 6 = 12 units

Step 2 Use absolute value to find the length of a horizontal side.

The endpoints of one horizontal side are (−5, 6) and (3, 6).

The x-coordinates of the endpoints are −5 and 3.

$|-5| = 5$ $|3| = 3$

The total length of the horizontal side is: 5 + 3 = 8 units

Step 3 Find the perimeter.

Opposite sides in a rectangle are the same length, so add the lengths of all four sides to find the perimeter, P.

P = 12 + 8 + 12 + 8 = 40 units

Each unit on the coordinate plane stands for 1 yard, so the actual perimeter is 40 yards.

Solution **The perimeter of the garden is 40 yards.**

Coached Example

Point *J* below at (4, 3) can be reflected across one or more axes to form a point at (−4, −3). Describe the reflection(s) that are needed.

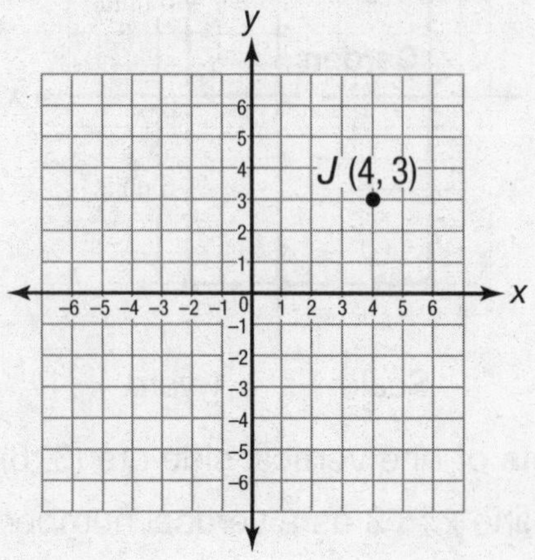

Compare the signs of the coordinates in (4, 3) and (−4, −3).

The *x*-coordinates of the ordered pairs have different signs.

The ____-coordinates of the ordered pairs have different signs, too.

Would point *J* need to be reflected across the *x*-axis, the *y*-axis, or both axes to move to (−4, −3) on the coordinate plane?

Since the signs of both coordinates are different, point *J* would need to be reflected across ________________.

Check your answer by reflecting (4, 3) across the *x*-axis only.

The coordinates of that reflected point are (4, ____).

Plot that point on the grid above and label it point *K*.

Now, reflect point *K* across the *y*-axis.

The coordinates of that reflected point are (____, ____).

Plot that point above and label it point *L*.

Point *L* is at (____, ____). Both of its coordinates have different signs than the coordinates of point *J*. Point *L* is a reflection of point *J* across ________________.

Lesson Practice

Choose the correct answer.

Use the coordinate plane for questions 1 and 2.

This coordinate plane shows the locations of 3 basketball team members during a practice drill.

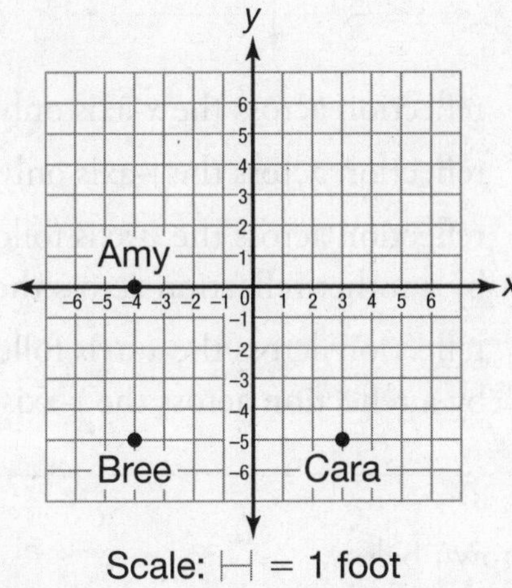

Scale: ⊢⊣ = 1 foot

1. Amy has the ball and throws it directly to Bree. How many feet does Amy throw the ball?

A. 6 feet

B. 5 feet

C. 4 feet

D. 0 feet

2. When Bree has the ball, she throws it directly to Cara. How many feet does Bree throw the ball?

A. 3 feet

B. 4 feet

C. 7 feet

D. 8 feet

Use the coordinate plane for questions 3 and 4.

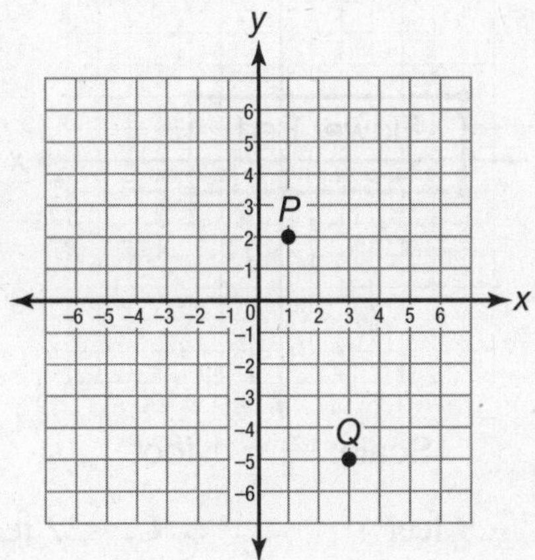

3. If point *P* is reflected over the *y*-axis, what will be the coordinates of its new location?

A. (2, 1)

B. (1, −2)

C. (−1, 2)

D. (−1, −2)

4. Which describes the new location of point *Q* after a reflection across the *x*-axis?

A. It will be at (3, 5).

B. It will be at (−3, 5).

C. It will be at (−3, −5).

D. It will be at the same location as point *P*.

5. The coordinate plane below shows a rectangular flower bed that Lucas will build. What will be the perimeter of the flower bed?

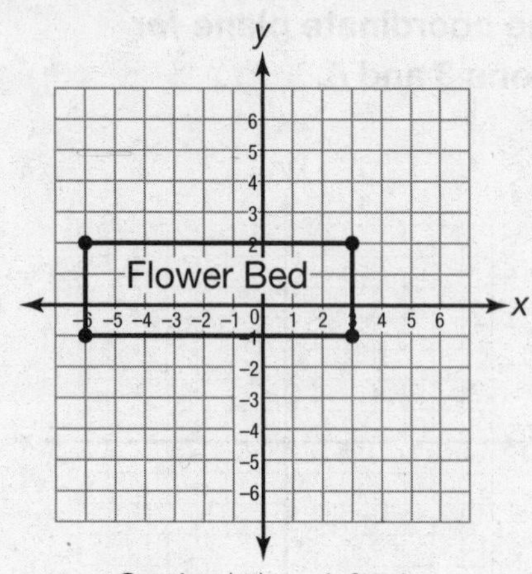

A. 9 feet
B. 12 feet
C. 22 feet
D. 24 feet

6. Point M below will be reflected so that it completely covers the point at $(5, -1)$. Which could describe this reflection?

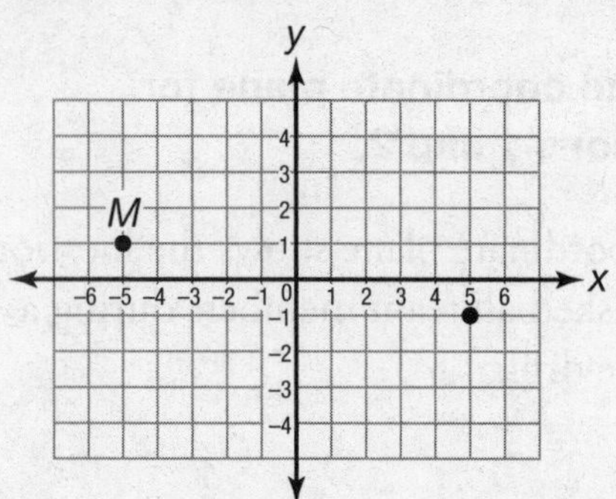

A. reflection across the x-axis only
B. reflection across the y-axis only
C. reflection across the x-axis followed by another reflection across the x-axis
D. reflection across the x-axis followed by a reflection across the y-axis

7. Mr. Chen drew a design for a rectangular sandbox, as shown below.

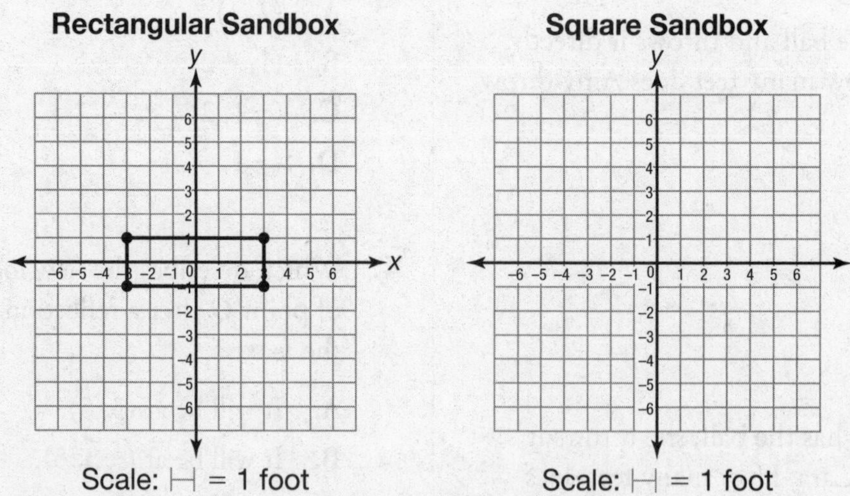

A. What will be the perimeter of the rectangular sandbox? ____________

B. On the coordinate plane above, draw a square sandbox with the same perimeter as the rectangular sandbox.

Domain 1: Cumulative Assessment for Lessons 1–11

1. Which situation would you describe with a negative integer?

A. a price increase of \$5

B. a 10-yard gain in football

C. a fall of 25 feet

D. a helicopter at 200 feet above a landing pad

2. Which is the opposite of -21?

A. -21

B. -12

C. 12

D. 21

3. Which is equivalent to $|-20|$?

A. -20

B. -2

C. 2

D. 20

4. Which point on the number line represents -0.4?

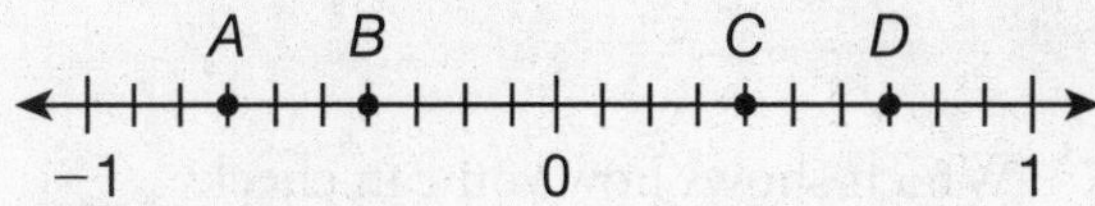

A. point A

B. point B

C. point C

D. point D

5. Which expression is equivalent to $28 + 35$?

A. $28(1 + 7)$

B. $7(4 + 5)$

C. $7(4 + 7)$

D. $4(7 + 9)$

6. If the numbers below were ordered from least to greatest, which number could you use to replace the $\square$?

$$\frac{1}{8}, \square, \frac{3}{10}, 0.6, \frac{3}{4}$$

A. $\frac{1}{5}$

B. 0.33

C. $\frac{2}{5}$

D. 0.8

7. What is 13,872 ÷ 34?

A. 48

B. 408

C. 480

D. 4,008

8. Which shows how you can check that $\frac{2}{3} \div \frac{7}{8} = \frac{16}{21}$?

A. $\frac{7}{8} \times \frac{16}{21} = \frac{2}{3}$

B. $\frac{16}{21} \div \frac{7}{8} = \frac{2}{3}$

C. $\frac{16}{21} \div \frac{2}{3} = \frac{7}{8}$

D. $\frac{8}{7} \times \frac{16}{21} = \frac{2}{3}$

9. Plot and label point P at (6, −1) on the coordinate grid.

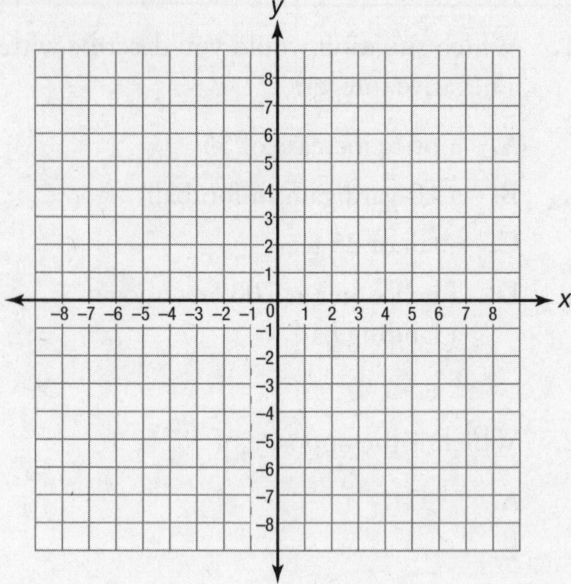

10. Mr. Harris is packaging items to give to his students. He has 48 pencils and 30 notebooks. He wants each package to contain the same number of pencils and the same number of notebooks.

A. What is the greatest number of packages Mr. Harris can make without having any items left over?

____

B. Explain how you found your answer for part A.

____

____

____

Domain 2

Ratios and Proportional Relationships

Domain 2: Diagnostic Assessment for Lessons 12–16

Use the diagram for questions 1 and 2.

1. In simplest form, what is the ratio of circles to triangles?

   A. 2:3
   B. 3:4
   C. 3:2
   D. 2:1

2. Which is **not** another way to express the ratio of all figures to squares?

   A. 9 to 2
   B. 9:2
   C. $\frac{9}{2}$
   D. $4\frac{1}{2}$

3. 147 is 35% of what number?

   A. 51.45
   B. 420
   C. 514.5
   D. 4,200

4. Which value of *x* will make these ratios equivalent?

   $$\frac{20}{35} = \frac{4}{x}$$

   A. 19
   B. 15
   C. 7
   D. 5

5. An artist mixes 2 parts blue paint and 3 parts yellow paint to get a shade of green. How many parts blue paint should the artist mix with 12 parts yellow paint to get the same shade of green?

   A. 6
   B. 8
   C. 9
   D. 13

6. Which of the following does **not** have a unit rate of $16 for one pair of pants?

   A. $32 for 2 pair of pants
   B. $48 for 3 pair of pants
   C. $64 for 4 pair of pants
   D. $90 for 5 pair of pants

7. A recipe for apple crisp uses 2 parts oats, 4 parts brown sugar, and 6 parts flour. In simplest form, how many parts of brown sugar are there for every part of flour?

   A. $\frac{2}{3}$ part
   B. $\frac{1}{2}$ part
   C. $\frac{1}{3}$ part
   D. $\frac{1}{4}$ part

8. Bart drove 140 miles in $2\frac{1}{2}$ hours. What was Bart's average speed in miles per hour?

   A. 66 miles per hour
   B. 65 miles per hour
   C. 56 miles per hour
   D. 55 miles per hour

9. The local movie theater collected a total of $8,250 last night, of which 45% came from sales at the concession stand. How much money did the theater collect at the concession stand?

   ____

10. The table shows the number of pounds of sugar and cocoa beans that a company uses to make chocolate candy bars.

Chocolate Candy Bars

Pounds of Sugar	3	6	9	12
Pounds of Cocoa Beans	2	4	6	?

A. How many pounds of cocoa beans should be mixed with 12 pounds of sugar?

____

B. Plot the ordered pairs from the table on the coordinate grid.

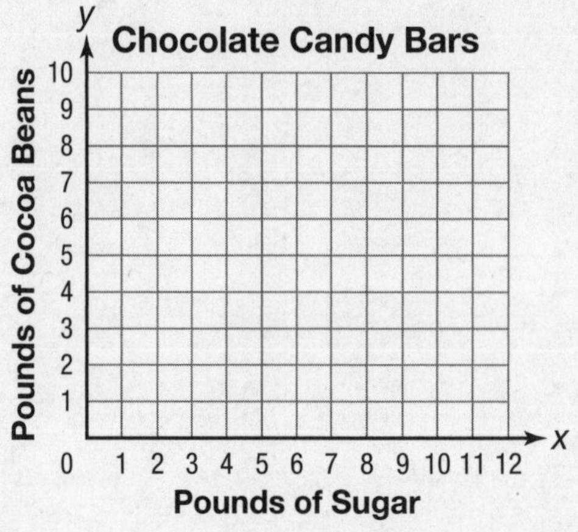

Common Core State Standard:
6.RP.1

Ratios

Getting the Idea

A **ratio** is a comparison of two numbers using division. Ratios are used to show relationships between quantities and can be written in three ways.

As a fraction	With a colon	In words
$\frac{a}{b}$	*a*:*b*	*a* to *b*

You can show different kinds of comparisons with a ratio.

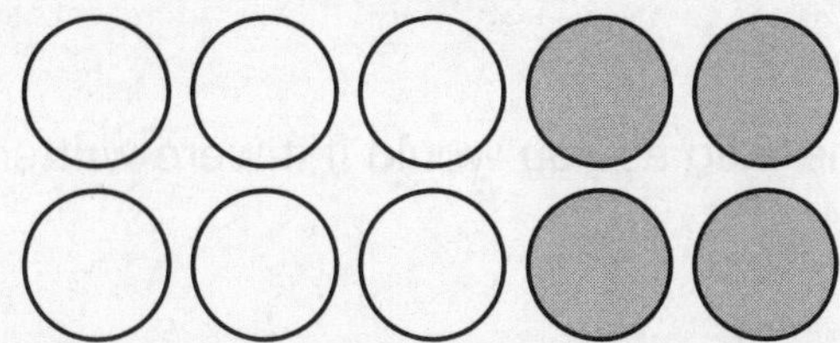

- Part to part: The ratio of white dots to gray dots is $\frac{6}{4}$, 6:4, or 6 to 4.
- Part to whole: The ratio of white dots to the total number of dots is $\frac{6}{10}$, 6:10, or 6 to 10.
- Whole to whole: The ratio of the number of triangles to the number of dots is $\frac{0}{10}$, 0:10, or 0 to 10.

As with fractions, you can simplify ratios. For example, the ratio of white dots to gray dots in the diagram above can be simplified as $\frac{3}{2}$, 3:2, or 3 to 2. A ratio should not be written as a mixed number, so do not simplify $\frac{3}{2}$ as $1\frac{1}{2}$.

Example 1

What is the ratio of cats to dogs in a neighborhood that has 9 dogs and 3 cats?

Strategy **Write a ratio to compare the number of cats to dogs.**

Step 1 Identify the numbers of cats and dogs.

There are 3 cats and 9 dogs.

Step 2 Write the ratio of cats to dogs.

The ratio of cats to dogs is $\frac{3}{9}$.

Step 3 Simplify.

$$\frac{3}{9} = \frac{3 \div 3}{9 \div 3} = \frac{1}{3}$$

Solution **The ratio of cats to dogs in the neighborhood is $\frac{1}{3}$, or there is 1 cat for every 3 dogs.**

Example 2

What is the ratio of girls to boys in a class of 20 students that has 12 boys? Write the ratio using a colon.

Strategy **Write a ratio to compare the number of girls to boys.**

Step 1 Find the numbers of boys and girls in the class.

There are 12 boys in the class.

$20 - 12 = 8$

There are 8 girls in the class.

Step 2 Write the ratio of girls to boys.

The ratio of girls to boys is 8:12.

Step 3 Simplify.

Simplify the ratio as you would if it were written in fraction form.

$\frac{8}{12} = \frac{8 \div 4}{12 \div 4} = \frac{2}{3}$

So, 8:12 = 2:3.

Solution **The ratio of girls to boys in the class is 2:3. There are 2 girls for every 3 boys.**

Example 3

The table shows the eye colors of the students in Mr. Matthew's class.

Eye Colors of Students

Color	Number of Students
Brown	15
Blue	6
Green	3
Hazel	2

For every student with green eyes, how many students have brown eyes?

Strategy **Find the ratio in simplest form.**

Step 1 Find the ratio of students with green eyes to students with brown eyes.

There are 3 students with green eyes and 15 students with brown eyes.

The ratio of students with green eyes to students with brown eyes is $\frac{3}{15}$.

Step 2 Simplify.

$\frac{3}{15} = \frac{3 \div 3}{15 \div 3} = \frac{1}{5}$

Solution **For every student with green eyes, there are 5 students with brown eyes.**

Coached Example

Matt's movie DVD collection contains 8 comedies, 6 dramas, 3 adventures, and 7 science fictions. What is the ratio of comedies and dramas to the total number of DVDs in Matt's collection? Write the ratio using the word *to*.

The ratio compares the total number of ______________ and ______________ to the total number of __________.

There are _____ comedies and _____ dramas.

Add: _____ + _____ = _____

There are _____ + _____ + _____ + _____ = _____ DVDs in all.

Write the ratio. _____ to _____

Simplify the ratio.

$\frac{14}{24} = \frac{14 \div \_\_}{24 \div \_\_} = \_\_\_\_$

So, _____ to _____ can be simplified as _____ to _____.

The ratio of comedies and dramas to total DVDs is ____________.

Lesson Practice

Choose the correct answer.

Use the diagram for questions 1 and 2.

1. What is the ratio of stars to hearts?

   A. 2:3
   B. 2:5
   C. 3:2
   D. 3:5

2. What is the ratio of all figures to stars?

   A. 5 to 2
   B. 5 to 3
   C. 3 to 5
   D. 2 to 3

3. Which is **not** another way to express the ratio 60 to 25?

   A. $\frac{12}{5}$
   B. 12:5
   C. 12 to 5
   D. $2\frac{2}{5}$

4. Paul has 16 baseballs and 12 golf balls. What is the ratio of baseballs to golf balls?

   A. $\frac{4}{7}$
   B. $\frac{3}{4}$
   C. $\frac{4}{3}$
   D. $\frac{7}{4}$

5. There were 15 cars and 25 trucks in the parking garage. What is the ratio of cars to total vehicles?

   A. 3:8
   B. 3:5
   C. 5:3
   D. 8:3

6. In his last soccer game, Chris made 9 saves and let in 3 goals. What was his ratio of saves to total shots on goal?

   A. $\frac{3}{1}$
   B. $\frac{4}{3}$
   C. $\frac{3}{4}$
   D. $\frac{1}{3}$

Use the table for questions 7 and 8.

Chantal's Flower Bed

Type of Flower	Number of Bulbs
Daffodils	8
Hyacinths	10
Tulips	12

7. Which ratio compares the number of daffodil bulbs to the number of tulip bulbs?

A. 3:4

B. 3:2

C. 4:3

D. 2:3

8. For every 3 bulbs that were planted, how many hyacinth bulbs were planted?

A. 1

B. 2

C. 3

D. 5

9. The table shows the number of votes that the candidates for class president received.

Votes for Class President

Candidate	Number of Votes
Brooke	18
Derek	36
Aidan	54
Julianne	27
Leonard	45

A. For every vote that Brooke received, who received three times as many?

____

B. What is the ratio in simplest form of the number of votes Derek received to the total number of votes? Explain how you found your answer.

____

____

____

Common Core State Standard:
6.RP.3.a

Equivalent Ratios

Getting the Idea

Remember, a ratio can be written in simplest form. For example, $\frac{8}{6}$ or 8:6 can be written in simplest form as $\frac{4}{3}$ or 4:3. The ratios $\frac{8}{6}$ and $\frac{4}{3}$ are **equivalent ratios.**

Models can help you tell if two ratios are equivalent or not.

Example 1

Use models to determine if the ratios below are equivalent.

3:4 and 9:12

Strategy **Model each ratio. Then try to separate 9:12 into groups of 3:4.**

Step 1 Model 3:4 using white dots and gray dots.

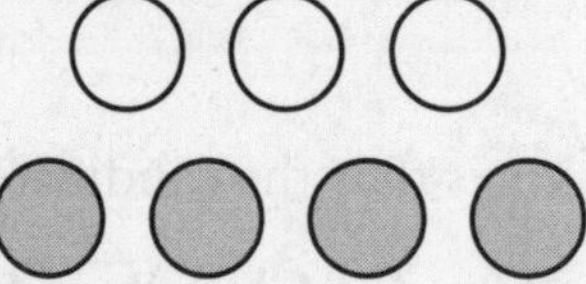

Step 2 Model 9:12 using white dots and gray dots.

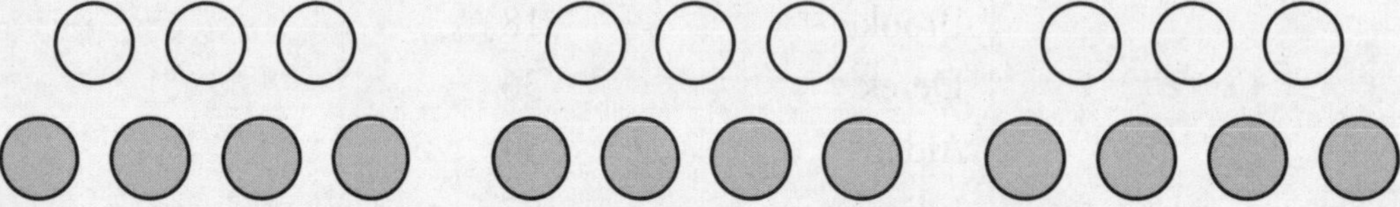

Step 3 Try to separate the model for 9:12 into groups of 3:4.

Circle groups of 3 white dots and 4 gray dots.

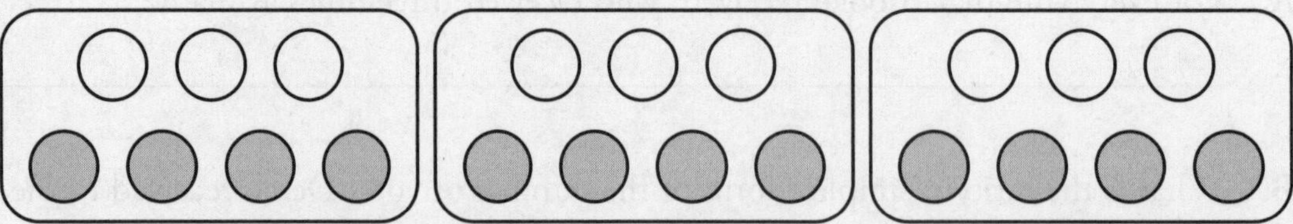

There are no ungrouped dots, so 3:4 and 9:12 are equivalent ratios.

Solution **The ratios 3:4 and 9:12 are equivalent.**

Another way to tell whether two ratios or rates are equivalent is to **cross multiply**. A rate is a ratio that compares two quantities with different units of measure. To cross multiply, write the ratios as fractions. Then multiply the numerator of each ratio and the denominator of the other ratio. The products are called the **cross products** of the ratios. If the cross products are equal, the ratios are equivalent.

Example 2

There are two photocopy machines in an office. The first machine produces 5 copies in 8 seconds. The second machine produces 16 copies in 24 minutes. Are the two machines making copies at the same rate?

Strategy **Express each rate as a fraction. Then cross multiply to see if the rates are equivalent.**

Step 1 Express each rate as a fraction.

$$\frac{5 \text{ copies}}{8 \text{ minutes}} = \frac{5}{8}$$

$$\frac{16 \text{ copies}}{24 \text{ minutes}} = \frac{16}{24}$$

Step 2 Cross multiply to see if the rates are equivalent.

$$\frac{5}{8} \stackrel{?}{=} \frac{16}{24}$$

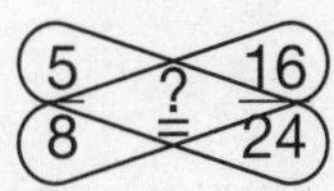

$$5 \times 24 \stackrel{?}{=} 8 \times 16$$

$$120 \neq 128$$

The cross products are not equal, so the ratios are not equivalent.

Solution **The two machines are not making copies at the same rate.**

You can cross multiply to find a missing value in a ratio or rate.

Example 3

Emma is driving at a constant speed. She drives 4 miles in 5 minutes. If she continues to drive at the same speed, how many miles will she drive in 15 minutes?

Strategy **Write two ratios to represent the situation. Cross multiply to find the missing term.**

Step 1 Write two ratios.

The first ratio is: $\frac{4 \text{ miles}}{5 \text{ minutes}}$ or $\frac{4}{5}$.

You want to find the number of miles she will drive in 15 minutes.

The second ratio is: $\frac{x \text{ miles}}{5 \text{ minutes}}$ or $\frac{x}{15}$.

Step 2 Cross multiply to find the missing term.

$$\frac{4}{5} = \frac{x}{15}$$
$$4 \times 15 = 5 \times x$$
$$60 = 5x$$
$$\frac{60}{5} = \frac{5x}{5}$$
$$12 = x$$

Solution **Emma will drive 12 miles in 15 minutes.**

You could also use a double number line to solve Example 3. The double number line shows the distances and times produced by a constant speed.

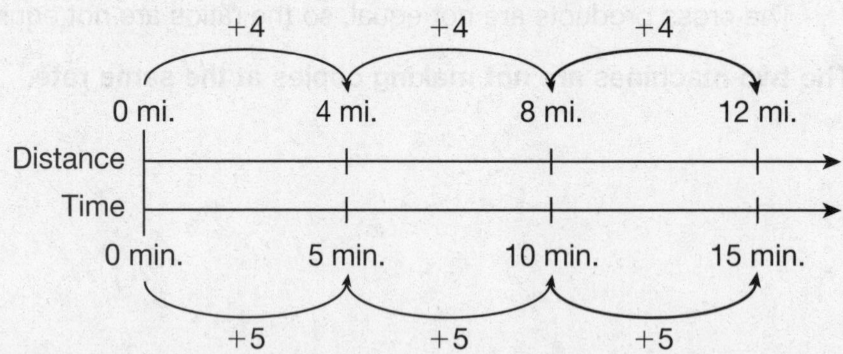

The double number line above shows that 4 miles in 5 minutes and 12 miles in 15 minutes are equivalent rates of speed.

You can look for patterns in a table to find equivalent ratios.

Example 4

A baker made this table to show the number of cups of flour and the number of eggs he needs to make a carrot cake recipe.

Carrot Cake Recipe

Cups of Flour (x)	2	3	4	5	6
Number of Eggs (y)	4	6	8	10	?

How many eggs does the baker need if he uses 6 cups of flour?

Strategy **Look for a rule that relates the number of cups of flour and the number of eggs. Then use that rule to write an equivalent ratio.**

Step 1 Look for a pattern that relates each pair of values in the table.

The first ratio is $\frac{\text{2 cups of flour}}{\text{4 eggs}}$, or $\frac{2}{4}$.

$2 \times \mathbf{2} = 4$, so the rule may be to multiply each x-value by 2.

See if that works for the other ratios, $\frac{3}{6}$, $\frac{4}{8}$, and $\frac{5}{10}$.

$3 \times \mathbf{2} = 6$ ✓

$4 \times \mathbf{2} = 8$ ✓

$5 \times \mathbf{2} = 10$ ✓

Step 2 Write a rule.

The number of eggs (y) equals 2 times the number of cups of flour (x).

The rule is: $y = 2 \times x$ or $y = 2x$.

Step 3 Find the missing value in the table.

If the baker uses 6 cups of flour, $x = 6$.

$y = 2x = 2 \times 6 = 12$

The ratio is 6 cups of flour to 12 eggs, or $\frac{6}{12}$.

The ratio $\frac{6}{12}$ in simplest form equals $\frac{1}{2}$, so those ratios are equivalent.

Solution **If the baker uses 6 cups of flour, he needs 12 eggs.**

You can also plot ordered pairs, (x, y), on a coordinate grid to find equivalent ratios. If the ratios of x:y are equivalent, the points will lie along the same straight line.

Example 5

The table shows the cost of shipping a package for different package weights.

Shipping Costs

Weight in Pounds (x)	1	2	4
Cost in Dollars (y)	3	6	12

Plot each ordered pair on a coordinate grid. Then use that graph to find two more equivalent rates.

Strategy **Plot the ordered pairs on a coordinate grid. Graph the line for the ordered pairs. Find two more ordered pairs on the line.**

Step 1 Plot points for each pair of values in the table.

Plot (1, 3), (2, 6), and (4, 12) on a coordinate grid.

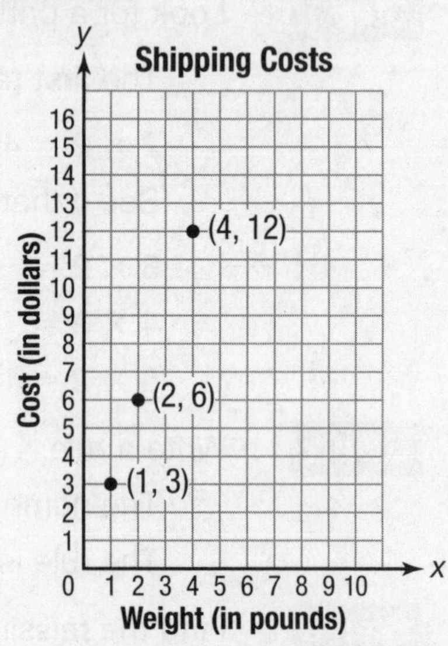

Step 2 Draw a line through the points. Extend the line. Name two other points on the line.

The line through points (1, 3), (2, 6), and (4, 12) shows that the ratios $\frac{1}{3}$, $\frac{2}{6}$, and $\frac{4}{12}$ are equivalent.

The points (3, 9) and (5, 15) are also on the line.

This shows that $\frac{3}{9}$ and $\frac{5}{15}$ are equivalent to the other ratios.

The points represent these rates: 3 pounds for \$9 and 5 pounds for \$15.

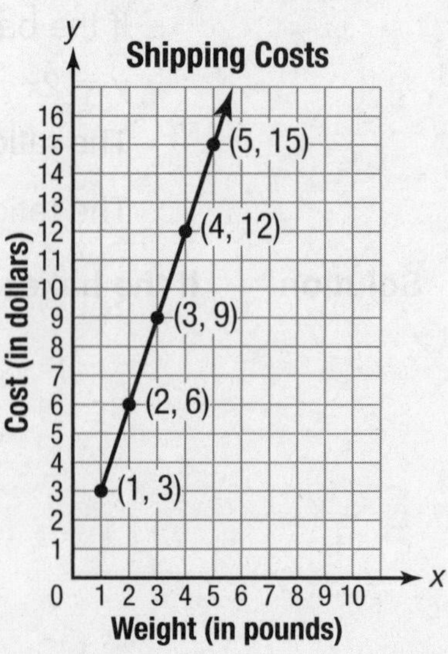

Solution **Two additional equivalent rates are 3 pounds for \$9 and 5 pounds for \$15.**

Coached Example

Hudson bought several cans of tennis balls. Each can contained both green and yellow tennis balls. He purchased 10 green tennis balls and 5 yellow tennis balls in all. If each can has 2 green tennis balls in it, how many yellow tennis balls are in each can?

Write two ratios for this problem.

Write the first ratio.

$$\frac{\text{10 green tennis balls}}{\text{\_\_\_\_ yellow tennis balls}} \text{ or } \frac{10}{\_\_\_\_}$$

You want to find the number of yellow tennis balls in a can of 2 green tennis balls.

Write the second ratio.

$$\frac{\text{2 green tennis balls}}{x \text{ yellow tennis balls}} \text{ or } \frac{2}{x}$$

Set the ratios equal to each other. Cross multiply to find the number of yellow tennis balls in each can.

$$\frac{10}{\_\_\_\_} = \frac{2}{x}$$

$$10 \times x = \_\_\_\_\_\_ \times 2$$

$$10x = \_\_\_\_\_\_$$

$$\frac{10x}{10} = \_\_\_\_\_\_$$

$$x = \_\_\_\_\_\_$$

Each can contains 2 green tennis balls and ______ yellow tennis ball(s).

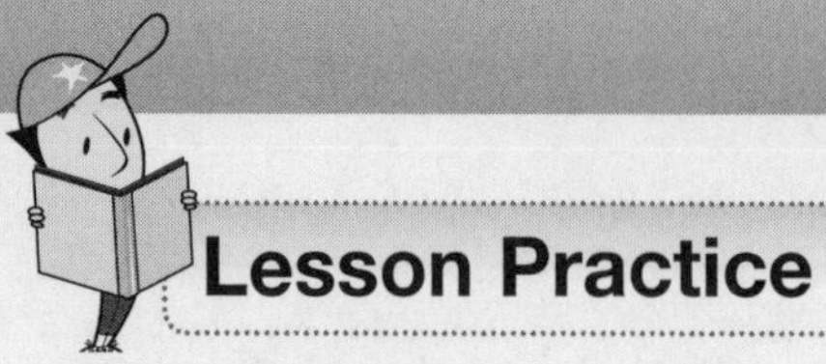

Choose the correct answer.

1. Which ratio is equivalent to $\frac{3}{10}$?

   A. $\frac{9}{10}$

   B. $\frac{9}{13}$

   C. $\frac{9}{20}$

   D. $\frac{9}{30}$

2. Which ratio is **not** equivalent to $\frac{5}{3}$?

   A. $\frac{35}{21}$

   B. $\frac{25}{15}$

   C. $\frac{18}{12}$

   D. $\frac{10}{6}$

3. Which pair of ratios are equivalent?

   A. $\frac{6}{9}$ and $\frac{12}{16}$

   B. $\frac{9}{15}$ and $\frac{18}{30}$

   C. $\frac{10}{18}$ and $\frac{16}{27}$

   D. $\frac{12}{15}$ and $\frac{15}{20}$

4. A television station shows 3 commercials every 12 minutes. At that rate, how many commercials will the station show in 60 minutes?

   A. 30

   B. 15

   C. 12

   D. 8

5. The table below shows the number of cups of sugar and of flour needed to make some cookies. If Alex uses 5 cups of sugar to make cookies, how many cups of flour does he need?

Cookie Ingredients

Cups of Flour	6	9	12	?
Cups of Sugar	2	3	4	5

   A. 20 cups

   B. 15 cups

   C. 13 cups

   D. 6 cups

6. The ratio of blue marbles to red marbles in a bag is 11:9. If there are 99 blue marbles in the bag, how many red marbles are there?

   A. 18

   B. 35

   C. 81

   D. 121

7. The ratio of boys to girls in a chorus is 5 to 6. Which shows an equivalent ratio?

   A. 10 boys to 12 girls

   B. 15 boys to 19 girls

   C. 20 boys to 25 girls

   D. 24 boys to 28 girls

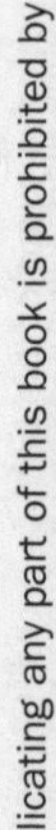

8. When biking at a constant speed, Abdul can travel 6 miles in 20 minutes. He made the double number line below to help him find how many miles he can bike in different amounts of time. How many miles can he bike in 40 minutes?

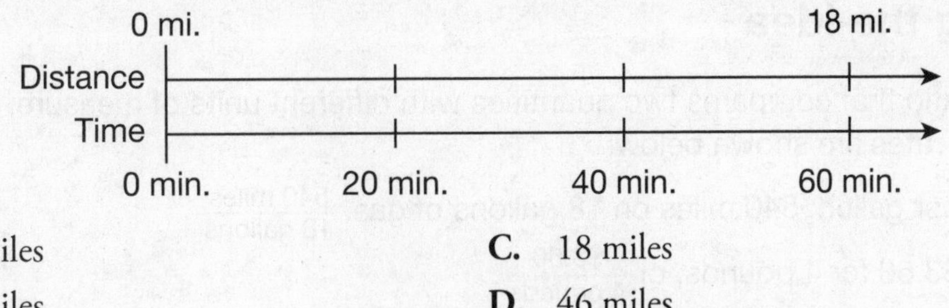

A. 2 miles
B. 12 miles
C. 18 miles
D. 46 miles

9. The table shows the number of cups of fruit juice and of ginger ale needed to make a fruit punch.

Fruit Punch

Cups of Fruit Juice (x)	2	4	6	8
Cups of Ginger Ale (y)	3	6	?	12

A. Do the pairs of values in the table represent equivalent ratios? Show your work or explain how you determined your answer.

B. Plot the ordered pairs from the table on the coordinate grid below. Then use the graph to determine how many cups of ginger ale must be mixed with 6 cups of fruit juice to make the punch.

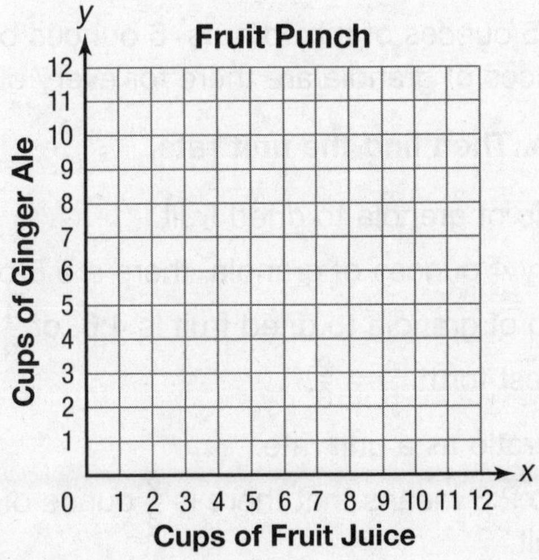

Common Core State Standards: 6.RP.2, 6.RP.3.b

Unit Rates

Getting the Idea

A **rate** is a ratio that compares two quantities with different units of measure. Some examples of rates are shown below:

- Miles per gallon: 540 miles on 18 gallons of gas, $\frac{540 \text{ miles}}{18 \text{ gallons}}$
- Cost: \$3.60 for 4 pounds, or $\frac{\$3.60}{4 \text{ pounds}}$
- Pay rate: \$285 for 30 hours, or $\frac{\$285}{30 \text{ hours}}$

Rates are often given as a **unit rate**, which is a rate in which the second measure is 1 unit. Each of the rates listed above can be simplified as unit rates.

- Miles per gallon: $\frac{540 \text{ miles}}{18 \text{ gallons}} = \frac{30 \text{ miles}}{1 \text{ gallon}}$
- Cost: $\frac{\$3.60}{4 \text{ pounds}} = \frac{\$0.90}{1 \text{ pound}}$
- Pay rate: $\frac{\$285}{30 \text{ hours}} = \frac{\$9.50}{1 \text{ hour}}$

In general, for every ratio a:b, the corresponding unit rate is $\frac{a}{b}$, where $b \neq 0$.

For example, if there are 4 cups of cranberry juice to every 5 cups of orange juice in a punch recipe, the ratio of cranberry juice to orange juice is 4:5, or $\frac{4}{5}$. That means that there is $\frac{4}{5}$ cup of cranberry juice for every 1 cup of orange juice. You can see this mathematically by multiplying each quantity by 5:

$$\frac{\frac{4}{5}}{1} = \frac{\frac{4}{5} \times 5}{1 \times 5} = \frac{4}{5}$$

Example 1

A recipe for trail mix uses 5 ounces of mixed nuts, 6 ounces of dried fruit, and 4 ounces of granola. How many ounces of granola are there for every ounce of dried fruit?

Strategy **Write a ratio. Then find the unit rate.**

Step 1 Write the ratio of granola to dried fruit.

For every 4 ounces of granola, there are 6 ounces of dried fruit.

The ratio of granola to dried fruit is 4:6, or $\frac{4}{6}$.

In simplest form, $\frac{4}{6} = \frac{2}{3}$.

Step 2 Interpret the ratio as a unit rate.

The ratio 2:3 means that there is $\frac{2}{3}$ ounce of granola for every ounce of dried fruit.

Step 3 Check your work.

Multiply by 6.

$$\frac{\frac{2}{3}}{1} = \frac{\frac{2}{3} \times 6}{1 \times 6} = \frac{4}{6}$$

For every 4 ounces of granola, there are 6 ounces of dried fruit.

Solution **There is $\frac{2}{3}$ ounce of granola for each ounce of dried fruit.**

To find a unit price, identify the quantities you want to compare and write a rate. Then simplify the rate to find the unit price.

Example 2

Mr. Wilson spent \$252 to stay 3 nights at Pavia Pavilions. At that rate, how much will he spend to stay 7 nights?

Strategy **Find the unit price. Then multiply by 7 nights.**

Step 1 Find the rate.

The rate is \$252 for 3 nights, or $\frac{252}{3}$.

Step 2 Find the unit rate, or unit price.

Divide 252 by 3 to find the price for one night.

$$\begin{array}{r} 84 \\ 3\overline{)252} \\ \underline{-24} \\ 12 \\ \underline{-12} \\ 0 \end{array}$$

The unit price is \$84 per night.

Step 3 Multiply the unit price by 7.

$7 \times 84 = 588$

Solution **Mr. Wilson will spend \$588 to stay 7 nights at Pavia Pavilions.**

In Example 2, you could also have set up equivalent ratios to solve the problem. Let x represent the cost of staying 7 nights.

$$\frac{252}{3} = \frac{x}{7}$$

$3 \times x = 252 \times 7$ ← Cross multiply.

$3x = 1{,}764$ ← Divide both sides by 3.

$x = 588$

A common use of rate is the speed formula $r = \frac{d}{t}$, or rate $= \frac{\text{distance}}{\text{time}}$.

Example 3

A train is traveling at a constant speed of 45 miles per hour. How far will the train travel in 2.5 hours?

Strategy **Use the speed formula.**

Step 1 Substitute the known values in the speed formula.

$$r = \frac{d}{t}$$

$$45 = \frac{d}{2.5} \text{ or } \frac{45}{1} = \frac{d}{2.5}$$

Step 2 Find an equivalent fraction for $\frac{45}{1}$ with a denominator of 2.5.

$$\frac{45}{1} = \frac{45 \times 2.5}{1 \times 2.5} = \frac{112.5}{2.5}$$

$$\frac{45 \text{ miles}}{1 \text{ hour}} = \frac{112.5 \text{ miles}}{2.5 \text{ hours}}$$

Solution **The train will travel 112.5 miles in 2.5 hours.**

You can rewrite the speed formula $r = \frac{d}{t}$ to solve for either distance, d, or time, t.

If $r = \frac{d}{t}$, then $d = r \times t$.

If $r = \frac{d}{t}$, then $t = \frac{d}{r}$.

In Example 3, you could have used the formula $d = r \times t$ to solve the problem.

$d = r \times t$

$d = 45 \times 2.5$

$d = 112.5$

Coached Example

Tanya walked 15 laps on an indoor track in 30 minutes. What was Tanya's average speed in laps per minute?

The speed formula is $r =$ ______.

The distance is ______ laps.

The time is ______ minutes.

Substitute the known values into the speed formula.

$r =$ ______.

Simplify the fraction.

$r =$ ______

Tanya's average speed was ______ laps per minute.

Lesson Practice

Choose the correct answer.

1. Which of the following is **not** an example of a rate?

   A. 2 cups for every 3 cups
   B. 120 beats per minute
   C. 16 ounces for $2
   D. 8 inches per 12 hours

2. Ling is driving at a constant speed of 55 miles per hour. At that rate, how long will it take him to drive 275 miles?

   A. 4 hours
   B. 5 hours
   C. 6 hours
   D. 7 hours

3. Callie's family spends an average of $70 per month on electricity. At that rate, what can Callie's family expect to pay for electricity over 1 year?

   A. $70
   B. $480
   C. $700
   D. $840

4. Mandy is on a bus that is traveling at a constant speed of 60 miles per hour. How far will she travel in $3\frac{1}{2}$ hours?

   A. 185 miles
   B. 195 miles
   C. 210 miles
   D. 230 miles

5. A party mix has 8 ounces of pretzels, 3 ounces of mini marshmallows, and 6 ounces of nuts. How many ounces of nuts are there for every ounce of pretzels?

   A. $\frac{6}{17}$ ounce of nuts for 1 ounce of pretzels
   B. $\frac{3}{8}$ ounce of nuts for 1 ounce of pretzels
   C. $\frac{1}{2}$ ounce of nuts for 1 ounce of pretzels
   D. $\frac{3}{4}$ ounce of nuts for 1 ounce of pretzels

6. Nate biked 54 miles in $4\frac{1}{2}$ hours. What was Nate's average speed in miles per hour?

   A. 11 miles per hour
   B. 12 miles per hour
   C. 13 miles per hour
   D. 14 miles per hour

7. Which of the following does **not** have a unit price of $24 for one sweater?

A. $38 for 2 sweaters

B. $72 for 3 sweaters

C. $96 for 4 sweaters

D. $120 for 5 sweaters

8. Fred's car can travel 368 miles on one tank of gas. His gas tank holds 16 gallons. What is the unit rate for miles per gallon?

A. 22 miles per gallon

B. 23 miles per gallon

C. 26 miles per gallon

D. 28 miles per gallon

9. A soup recipe uses 6 cups of water, 4 cups of tomato sauce, and 5 cups of tomato puree.

A. How many cups of tomato sauce are there for every cup of water?

____

B. If 4 cans of tomato sauce cost $2, what is the unit price for 1 can of tomato sauce? Show your work.

____

Common Core State Standard:
6.RP.3.c

Percents

Getting the Idea

A **percent** is a ratio that means "per hundred." The symbol for percent is %. A percent can be written as a fraction with a denominator of 100 or as a decimal. To write a percent as a decimal, divide the percent by 100 and remove the percent sign. Dividing by 100 is the same as moving the decimal point two places to the left. For example, $45\% = \frac{45}{100} = 0.45$.

Example 1

What fraction, percent, and decimal name the shaded part of this grid?

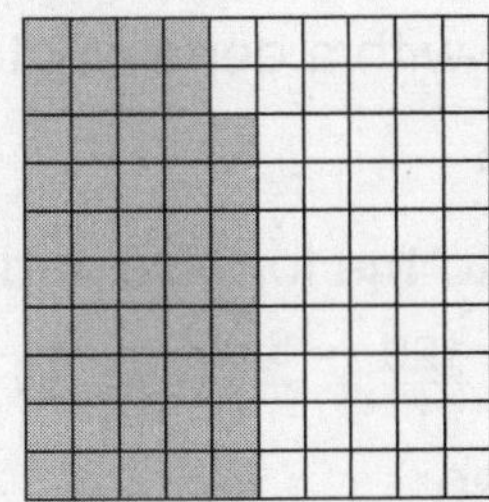

Strategy **Write the fraction of the grid that is shaded. Then convert the fraction to a percent and a decimal.**

Step 1 Write the fraction.

Count the number of shaded squares.

There are 48 shaded squares. There are a total of 100 squares.

$\frac{48}{100}$ of the grid is shaded.

Step 2 Write the percent.

Write the numerator of the fraction with a percent symbol after it.

$\frac{48}{100} = 48\%$

Step 3 Write the decimal.

Divide the percent by 100 and drop the percent sign.

$48\% \div 100 = 0.48$

Solution $\frac{48}{100}$**, 48%, and 0.48 name the shaded part of the grid.**

You can use multiplication to solve percent problems. When you multiply a number by a percent, you are finding a part of a whole. The formula below will help you.

percent $\times$ whole $=$ part

Example 2

What is 25% of 120?

Strategy **Find a part of a whole using the formula percent $\times$ whole $=$ part.**

Step 1 Identify the known and unknown values in the formula.

The percent is 25%.

The whole is 120.

The part is unknown.

Step 2 Write 25% as a fraction with a denominator of 100.

$$25\% = \frac{25}{100}$$

Step 3 Substitute the values into the formula and solve.

$$\frac{25}{100} \times 120 = \frac{25}{100} \times \frac{120}{1} = \frac{3{,}000}{100} = 30$$

So, 30 is 25% of 120.

Solution **25% of 120 is 30.**

Example 3

There are 180 students going on a class field trip. Of those students, 40% are boys. How many students on the field trip are boys?

Strategy **Write the percent as a fraction. Multiply.**

Step 1 Write the percent as a fraction in simplest form.

$$40\% = \frac{40 \div 20}{100 \div 20} = \frac{2}{5}$$

Step 2 Use the formula: percent $\times$ whole $=$ part.

Multiply the total number of students by the fraction.

$$\frac{2}{5} \times 180 = \frac{2}{\cancel{5}_{1}} \times \frac{\cancel{180}^{36}}{1} = 72$$

Solution **There are 72 boys on the field trip.**

If the part and the percent are known, you can rewrite the formula to find the whole.

If percent × whole = part, then whole = part ÷ percent.

Example 4

A group of students is trying out for the soccer team. Of those students, 22 are seventh graders. If 55% of the students trying out are seventh graders, how many students in all are trying out?

Strategy **Find the whole using the formula: whole = part ÷ percent.**

Step 1 Rename the percent as a decimal.

$55\% = \frac{55}{100} = 0.55$

Step 2 Substitute known values into the formula.

whole = part ÷ percent

whole = 22 ÷ 0.55

whole = 40

Solution **There are a total of 40 students trying out for the soccer team.**

Coached Example

A pet store has 52 freshwater fish. Of all the fish in the store, 80% are freshwater fish. How many fish does the store have in all?

To find the total number of fish, use the formula for finding the whole.

whole = part ÷ percent

Rename the percent as a decimal.

$80\% = \frac{\_\_\_}{100} = \_\_\_\_\_\_$

Substitute known values into the formula.

whole = ______ ÷ ______ = ______

The store has ______ fish in all.

Lesson Practice

Choose the correct answer.

1. What percent of the grid is shaded?

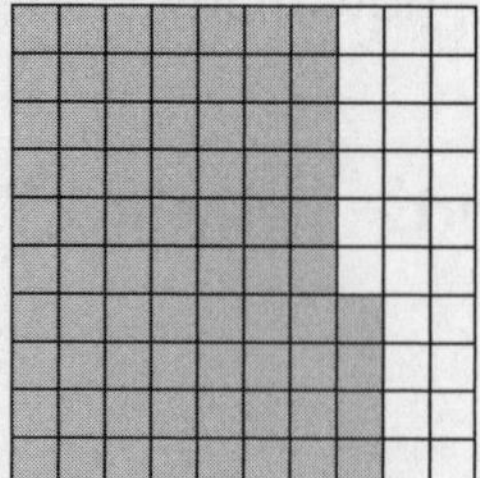

A. 0.74%

B. 7.4%

C. 74%

D. 740%

2. Which fraction is equal to 30%?

A. $\frac{3}{100}$

B. $\frac{1}{30}$

C. $\frac{3}{10}$

D. $\frac{1}{3}$

3. Which decimal is equal to 42%?

A. 0.42

B. 4.2

C. 42

D. 420

4. There are 25 students performing in the holiday concert. Of the students, 11 are boys. What percent of the students are boys?

A. 44%

B. 48%

C. 52%

D. 56%

5. What is 38% of 560?

A. 21.28

B. 212.8

C. 213

D. 224

6. 435 is 15% of what number?

A. 6.525

B. 65.25

C. 2,900

D. 29,000

7. Cecilia created 2.5 liters of a chemical mixture for an experiment. Saline solution accounted for 12.5% of the mixture. How many liters of saline solution were in the mixture?

A. 0.3125 liter

B. 2.1875 liters

C. 3.125 liters

D. 31.25 liters

8. As part of a class project, Mesut collected $126 in donations for a local hospice. That amount was 7% of the total amount collected by the class. How much money did the class collect in all?

A. $167.40

B. $180

C. $1,674

D. $1,800

9. Last year, a state university received 3,560 applications from boys. Of those applications, 35% were from boys who lived in other states.

A. How many applications did the university receive from boys who lived in other states?

____

B. Applications to the university from boys represented 40% of all applications. How many applications did the university receive in all? Explain how you found your answer.

____

Common Core State Standard:
6.RP.3.d

Convert Measurements

Getting the Idea

The tables below show some conversions for units of **length** in both the **customary system** and the **metric system**.

Customary Units of Length

1 foot (ft) = 12 **inches** (in.)
1 **yard** (yd) = 3 feet
1 yard = 36 inches
1 **mile** (mi) = 5,280 feet
1 mile = 1,760 yards

Metric Units of Length

1 **centimeter** (cm) = 10 **millimeters** (mm)
1 **meter** (m) = 100 centimeters
1 meter = 1,000 millimeters
1 **kilometer** (km) = 1,000 meters

You can convert measurements using equivalent ratios.

Example 1

Nancy ran 8 miles. How many yards did she run?

Strategy **Set up equivalent ratios and cross multiply.**

Step 1 Write a ratio that compares yards to miles.

$$\frac{\text{yards}}{\text{miles}} = \frac{1,760}{1}$$

Step 2 Write a ratio that compares the unknown length to the length you know.

Let y represent the number of yards.

$$\frac{\text{yards}}{\text{miles}} = \frac{y}{8}$$

Step 3 Set up equivalent ratios using the two ratios.

$$\frac{1,760}{1} = \frac{y}{8}$$

Step 4 Cross multiply.

$$1,760 \times 8 = 1 \times y$$

$$14,080 = y$$

Solution **Nancy ran 14,080 yards.**

The tables below show conversions among units of **weight** and **mass**.

Customary Units of Weight

1 **pound** (lb) = 16 **ounces** (oz)
1 **ton** (T) = 2,000 pounds

Metric Units of Mass

1 **gram** (g) = 1,000 **milligrams** (mg)
1 **kilogram** (kg) = 1,000 grams
1 **metric ton** (t) = 1,000 kilograms

Example 2

How many grams are equal to 5 kilograms?

Strategy **Set up equivalent ratios and cross multiply.**

Step 1 Write a ratio that compares grams to kilograms.

$$\frac{\text{grams}}{\text{kilograms}} = \frac{1{,}000}{1}$$

Step 2 Write a ratio that compares the unknown mass to the mass you know.

Let g represent the number of grams.

$$\frac{\text{grams}}{\text{kilograms}} = \frac{g}{5}$$

Step 3 Set up equivalent ratios using the two ratios.

$$\frac{1{,}000}{1} = \frac{g}{5}$$

Step 4 Cross multiply.

$$1{,}000 \times 5 = 1 \times g$$

$$5{,}000 = g$$

Solution **There are 5,000 grams in 5 kilograms.**

Compound units can be used to express measurements. For example, you may express a weight in ounces, or you may express the same weight using pounds and ounces.

Example 3

A newborn baby weighed 133 ounces. What is the baby's weight in pounds and ounces?

Strategy **Set up equivalent ratios and cross multiply.**

Step 1 Write a ratio that compares ounces to pounds.

$$\frac{\text{ounces}}{\text{pounds}} = \frac{16}{1}$$

Step 2 Write a ratio that compares the unknown weight to the weight you know.

Let p represent the number of pounds.

$$\frac{\text{ounces}}{\text{pounds}} = \frac{133}{p}$$

Step 3 Set up equivalent ratios using the two ratios.

$$\frac{16}{1} = \frac{133}{p}$$

Step 4 Cross multiply.

$$16 \times p = 1 \times 133$$
$$16p = 133$$
$$16p \div 16 = 133 \div 16$$
$$p = 8 \text{ R}5$$

The remainder is the additional number of ounces. There are 5 ounces.

Solution **The newborn baby weighed 8 pounds 5 ounces.**

The tables below shows conversions among units of **capacity**.

Customary Units of Capacity

1 **cup** (c) = 8 **fluid ounces** (fl oz)
1 **pint** (pt) = 2 cups
1 **quart** (qt) = 2 pints
1 **gallon** (gal) = 4 quarts

Metric Units of Capacity

1 **liter** (L) = 1,000 **milliliters** (mL)

Example 4

A fishbowl has a capacity of 192 fluid ounces. How many quarts is that?

Strategy **Set up equivalent ratios and cross multiply.**

Step 1 Find the number of fluid ounces in a quart.

1 cup = 8 fluid ounces

2 cups = 16 fluid ounces = 1 pint

4 cups = 32 fluid ounces = 2 pints = 1 quart

There are 32 fluid ounces in a quart.

Step 2 Write a ratio that compares fluid ounces to quarts.

$$\frac{\text{fluid ounces}}{\text{quarts}} = \frac{32}{1}$$

Step 3 Write a ratio that compares the capacity you know to the unknown capacity.

Let q represent the number of quarts.

$$\frac{\text{fluid ounces}}{\text{quarts}} = \frac{192}{q}$$

Step 4 Set up equivalent ratios using the two ratios.

$$\frac{32}{1} = \frac{192}{q}$$

Step 5 Cross multiply.

$$32 \times q = 1 \times 192$$

$$32q = 192$$

$$32q \div 32 = 192 \div 32$$

$$q = 6$$

Solution **The capacity of the fishbowl is 6 quarts.**

Coached Example

A basketball player is 78 inches tall. What is his height in feet?

Write a ratio that compares inches to feet. Remember, 12 inches = 1 foot.

$\frac{\text{inches}}{\text{feet}} = \frac{\_\_\_\_}{1}$

Write a ratio that compares the height you know to the unknown height.

Let *f* represent the height in feet.

$\frac{\text{inches}}{\text{feet}} = \frac{\_\_\_\_}{f}$

Set up equivalent ratios using the two ratios.

_____ = _____

Solve. Show your work.

The remainder is the additional number of __________. The remainder is _____, so there are _____ additional inches.

1 foot = 12 inches, so 6 inches = _____ foot.

The basketball player is _____ feet tall.

Lesson Practice

Choose the correct answer.

1. Dean's car weighs $1\frac{1}{4}$ tons. How many pounds does his car weigh?

A. 2,000 lb
B. 2,125 lb
C. 2,375 lb
D. 2,500 lb

2. A wooden board is 3 yards 1 foot long. Which shows an equivalent length?

A. 37 ft
B. 31 ft
C. 10 ft
D. 7 ft

3. A bottle of water has a capacity of 750 milliliters. Which is an equivalent measure in liters?

A. 7,500 L
B. 75 L
C. 7.5 L
D. 0.75 L

4. A package weighs 4.25 kilograms. How many grams does the package weigh?

A. 0.0425 g
B. 425 g
C. 4,250 g
D. 42,500 g

5. Julie's cell phone is 9 centimeters long. How many millimeters long is her cell phone?

A. 0.9 millimeters
B. 90 millimeters
C. 900 millimeters
D. 9,000 millimeters

6. Amy needs to fill a barrel with $4\frac{1}{4}$ gallons of water. She only has a quart container. How many times will she need to fill the quart container in order to get $4\frac{1}{4}$ gallons of water into the barrel?

A. 34
B. 20
C. 17
D. 16

7. Michael will be running a 15-mile road race this weekend. How many feet will he run?

A. 26,400 feet
B. 26,700 feet
C. 78,900 feet
D. 79,200 feet

8. Melissa made 164 fluid ounces of lemonade for a party. How many cups of lemonade did Melissa make?

A. $10\frac{1}{4}$ cups
B. $20\frac{1}{4}$ cups
C. $20\frac{1}{2}$ cups
D. 21 cups

9. Brendan's dog has a mass of 25,700 grams.

A. What is the dog's mass in milligrams?

____

B. What is the dog's mass in kilograms? Explain how you found your answer.

____

Domain 2: Cumulative Assessment for Lessons 12–16

Use the diagram for questions 1 and 2.

1. In simplest form, what is the ratio of rectangles to stars?

A. 1:2

B. 4:5

C. 2:1

D. 5:2

2. Which is **not** another way to express the ratio of all figures to rectangles?

A. 11 to 4

B. 11:4

C. $\frac{11}{4}$

D. $2\frac{3}{4}$

3. 24 is 16% of what number?

A. 0.384

B. 15

C. 38.4

D. 150

4. Which value of x will make these ratios equivalent?

$$\frac{25}{30} = \frac{5}{x}$$

A. 20 **C.** 6

B. 10 **D.** 5

5. A recipe for 4 loaves of bread uses 3 tablespoons of honey. How much honey is needed for 24 loaves of bread?

A. 6 teaspoons

B. 8 teaspoons

C. 12 teaspoons

D. 18 teaspoons

6. Which of the following does **not** have a unit rate of $14 for one CD?

A. $28 for 2 CDs

B. $45 for 3 CDs

C. $56 for 4 CDs

D. $70 for 5 CDs

7. A banana cream pie recipe uses 8 ounces of cream cheese, 14 ounces of condensed milk, and 12 ounces whipped topping. In simplest form, how many ounces of cream cheese are there for every ounce of condensed milk?

A. $\frac{4}{7}$ ounce

B. $\frac{1}{4}$ ounce

C. $\frac{4}{13}$ ounce

D. $\frac{2}{3}$ ounce

8. Miley drove 288 miles in $4\frac{1}{2}$ hours. What was Miley's average speed in miles per hour?

A. 58 miles per hour

B. 52 miles per hour

C. 64 miles per hour

D. 66 miles per hour

9. The Shakespeare festival produced by a local theater company was attended by 14,350 people, of whom 28% were high school students. How many high school students attended the Shakespeare festival?

____

10. The table shows the teaspoons of salt and of sugar in a recipe.

Sugar and Salt in a Recipe

Teaspoons of Salt (x)	2	4	6	8
Teaspoons of Sugar (y)	3	6	9	?

A. How many parts of sugar should be used with 8 parts of salt?

B. Plot the ordered pairs from the table on the coordinate grid.

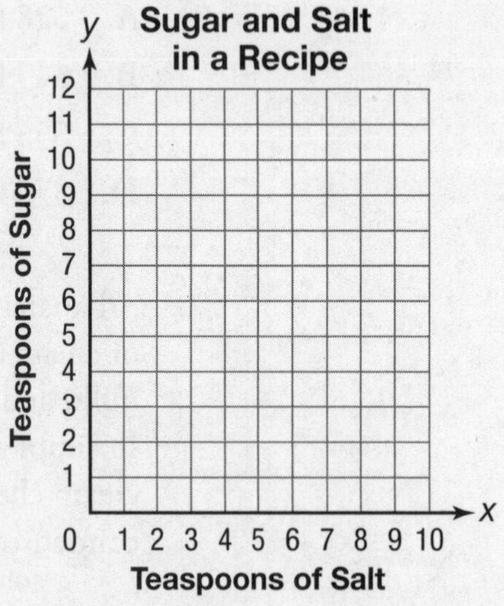

Domain 3

Expressions and Equations

Domain 3: Diagnostic Assessment for Lessons 17–23

1. Which expression represents "the product of 8 squared and the difference of a number n and 9"?

   A. $8^2 \times (n - 9)$

   B. $8^2 \times (n + 9)$

   C. $8^2 + 9n$

   D. $8^2 - 9n$

2. What is the value of the expression below when $a = 4$?

   $6a + 7$

   A. 31

   B. 41

   C. 53

   D. 71

3. Which expression is equivalent to $9(4 + r)$?

   A. $36r$

   B. $36 + r$

   C. $13r$

   D. $36 + 9r$

4. Which number is a solution for the inequality below?

   $5x \leq 20$

   A. 90

   B. 20

   C. 10

   D. 0

5. Franklin paid \$152 for 8 DVDs. Each DVD was the same price. Which shows the equation that represents the situation and the price of each DVD?

   A. $\frac{d}{8} = 152$; \$1,216

   B. $d - 8 = 152$; \$160

   C. $d + 8 = 152$; \$144

   D. $8d = 152$; \$19

6. Which equation best represents the relationship between x and y shown in the table?

x	0	1	2	3
y	1	4	7	10

   A. $y = x + 1$

   B. $y = 3x + 1$

   C. $y = 4x$

   D. $y = 5x - 1$

7. Josephine bought a CD that cost \$18. She handed the clerk d dollars. She received more than \$30 in change. Which inequality best represents d, the number of dollars she handed the clerk?

   A. $d > 30$

   B. $d \geq 30$

   C. $d > 48$

   D. $d \geq 48$

8. What is the value of k in the following equation?

$\frac{1}{4}k = 8$

A. 2
B. 16
C. 32
D. 64

9. Describe the expression $15 + (12 \div n)$ in words.

____

____

10. The equation $y = x - 2$ describes how the variables x and y are related.

A. Complete the table of values below for $y = x - 2$. Show all your work.

x	$y = x - 2$	y	(x, y)
2			
4			
6			
8			

B. Graph $y = x - 2$ on the coordinate grid below.

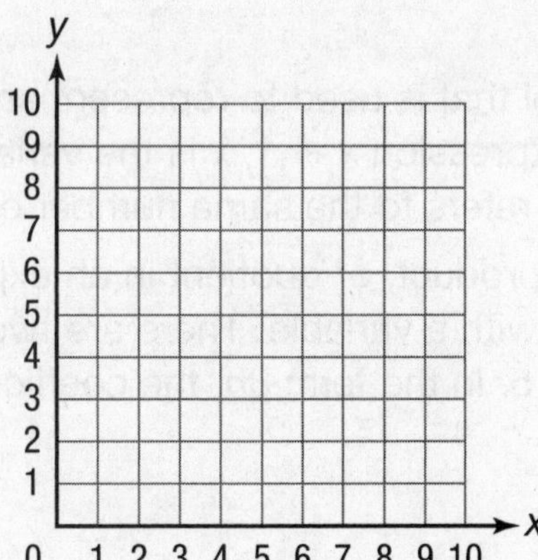

Write Expressions

Common Core State Standards:
6.EE.1, 6.EE.2.a, 6.EE.2.b, 6.EE.6

Getting the Idea

An **expression** is a mathematical phrase with numbers, operation signs, and variables. You can write an expression to describe a real-world situation. For example, the numerical expression $6 \div 3$ could describe placing 6 students into 3 equal groups.

Example 1

Jeremiah chose a card with a number on it. He wrote the expression 2^3 to describe the number on the card. What is the number on the card?

Strategy **Evaluate the meaning of the exponential expression.**

Step 1 An exponent of 3 means you use the base as a factor 3 times.

$2^3 = 2 \times 2 \times 2$

Step 2 Multiply.

$2 \times 2 \times 2 = 4 \times 2 = 8$

Solution **The number on the card is 8.**

A **variable** is a letter or symbol that is used to represent an unknown number in an algebraic expression. In the expression $x + 1$, x is the variable. If a variable occurs in an expression more than once, it refers to the same number each time.

A **term** is a number, variable, product, or quotient in an expression. A **coefficient** is the numerical factor in a term with a variable. There are two terms in the expression $3n + 5$. The terms are $3n$ and 5. In the term $3n$, the coefficient is 3.

Example 2

George had 5 apples. His mother gave him a few more apples to share with his friends. Write an expression to represent the total number of apples George has.

Strategy **Identify the key word. Choose a variable to represent the unknown number of apples.**

Step 1 Identify the key word.

The key word is "more." This means addition.

Step 2 Choose a variable.

Let a represent the number of apples George's mother gave him.

Step 3 Write the expression.

George already had 5 apples, and then he was given *a* apples by his mother.

George now has $5 + a$ apples.

Solution **George has a total of $5 + a$ apples.**

Example 3

Write an expression for this statement: four times as much as a number *k*.

Strategy **Identify the key word or words.**

Step 1 Identify the key word.

The key word is "times." This means multiplication.

Step 2 Identify the values being multiplied.

4 and *k*

Step 3 Write the expression using the operation and values.

$4 \times k$, or $4k$

Solution **An expression for "four times as much as a number *k*" is $4k$.**

Example 4

Michelle writes 15 pages in her journal each week, plus an extra 5 pages on her birthday. Michelle's birthday was this week. Write an expression to show how many pages Michelle has written this year.

Strategy **Use a variable to write an expression.**

Step 1 Choose a variable for the unknown number.

You do not know how many weeks have passed this year.

Let *w* equal the number of weeks Michelle has been writing.

Step 2 Write the expression.

$15w + 5$

Solution **Michelle has written $15w + 5$ journal pages this year.**

Example 5

Write an expression for this statement: the product of 8 and the difference of a number and 12.

Strategy **Identify the key word or words.**

Step 1 Identify the key words.

The key words are "product" and "difference." Product implies multiplication and difference implies subtraction.

Step 2 Identify the values being subtracted and write an expression.

You are looking for "the difference of a number and 12."

Let n represent the number.

$n - 12$

Step 3 Write the expression for the entire statement.

You are looking for the product of 8 and $n - 12$.

$8 \times (n - 12)$, or $8(n - 12)$

Solution **An expression for "the product of 8 and the difference of a number and 12" is $8(n - 12)$.**

Example 6

Write an expression for this statement: the quotient of 80 and 4 squared.

Strategy **Identify the key word or words.**

Step 1 Identify the key words.

The key words are "quotient" and "squared." Quotient means division and squared means an exponent.

Step 2 Write an expression for "4 squared."

"4 squared" can be written as 4^2.

Step 3 Write an expression for the entire statement.

You are looking for the quotient of 80 and 4^2.

$80 \div 4^2$

Solution **An expression for "the quotient of 80 and 4 squared" is $80 \div 4^2$.**

Coached Example

Write an expression that represents "twice a number *n* increased by 11."

The key words are __________ and __________.

"Twice a number" means to ______________ by __________.

"Increased by 11" means to ______________ 11.

Let *n* represent the unknown quantity.

Identify the terms to use in the expression: __________ and __________

Write an expression using the operations and terms. ______________

An expression that represents "twice a number *n* increased by 11" is ______.

Lesson Practice

Choose the correct answer.

1. Which expression represents "the product of a number g and 8"?

   A. $g + 8$
   B. $g - 8$
   C. $8g$
   D. $8 \div g$

2. Which expression represents "half the sum of 5 and a number b"?

   A. $\frac{b + 5}{2}$
   B. $2b + 5$
   C. $\frac{b}{2 + 5}$
   D. $\frac{b}{2} + 5$

3. Bianca puts \$10 in a savings account each month and an extra \$20 when she receives money for her birthday. If her birthday was this week, which expression represents the amount she has saved this year?

   A. $20m + 10$
   B. $20m - 10$
   C. $10m - 20$
   D. $10m + 20$

4. Marion is 3 years more than 5 times as old as Paula. If p represents Paula's age, which expression represents Marion's age?

   A. $3p + 5$
   B. $3p - 5$
   C. $5p + 3$
   D. $5p - 3$

5. Which expression represents "add 7 and a number n, then multiply by 8 cubed"?

   A. $8^3 \times (7 + n)$
   B. $8^3 + 7n$
   C. $8^3 + 7 + n$
   D. $8^3 \times 7n$

6. Oscar bought n ride tickets at the carnival. Esther bought 4 times as many ride tickets as Oscar. Which expression represents the total number of ride tickets that Oscar and Esther bought?

   A. $4n + 4n$
   B. $n + 4n$
   C. $n + 4$
   D. $4n$

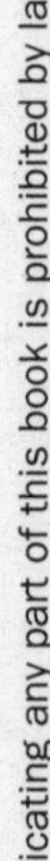

7. Which expression represents "9 less than the product of 5 and a number n"?

A. $9 - (5 + n)$

B. $9 - 5n$

C. $(5 + n) - 9$

D. $5n - 9$

8. Which expression represents "the sum of 16 squared and the quotient of 8 and a number b?

A. $16^2 + \frac{8}{b}$

B. $16^2 + 8b$

C. $(16 + 8)^2 \div b$

D. $(16 + 8 \div b)^2$

9. Use "the product of 6 and the sum of 3 times a number n and 5" to answer the questions below.

A. Write an expression that represents the statement.

____

B. Explain how you decided what operation symbols to use in your expression.

____

____

____

Common Core State Standards: 6.EE.1, 6.EE.2.c

Evaluate Expressions

Getting the Idea

You can evaluate an expression with a variable by substituting a number for the variable. If an expression contains more than one operation, you need to know in which order to perform the operations. The **order of operations** is a set of rules that determines the correct sequence for evaluating expressions.

Order of Operations

1. Evaluate expressions in parentheses.
2. Evaluate exponents.
3. Multiply and divide in order from left to right.
4. Add and subtract in order from left to right.

Example 1

Evaluate the expression $3^2 + 5$.

Strategy **Use the order of operations.**

Step 1 There are no parentheses, so evaluate exponents first.

$3^2 + 5 = 9 + 5$

Step 2 There is no multiplication or division, so add next.

$9 + 5 = 14$

Solution $\mathbf{3^2 + 5 = 14}$

Example 2

Evaluate $7x - 5$ for $x = 4$.

Strategy **Substitute the given value of *x* in the expression. Then evaluate.**

Step 1 Rewrite the expression, substituting 4 for *x*.

$7 \times 4 - 5$

Step 2 Evaluate the expression using the order of operations.

$7 \times 4 - 5$ ← Multiply first.

$28 - 5$ ← Then subtract.

23

Solution **When $x = 4$, $7x - 5 = 23$.**

Example 3

Evaluate $a^2 + 6$ for $a = 5$.

Strategy **Substitute the given value of *a* in the expression. Then evaluate.**

Step 1 Rewrite the expression, substituting 5 for *a*.

$5^2 + 6$

Step 2 Evaluate the expression using the order of operations.

$5^2 + 6$ ← Evaluate exponents first.

$25 + 6$ ← Then add.

31

Solution **When $a = 5$, $a^2 + 6 = 31$.**

Example 4

Evaluate $a^3 \div (2 \times b)$ for $a = 4$ and $b = 8$.

Strategy **Substitute the given values of *a* and *b* in the expression. Then evaluate.**

Step 1 Rewrite the expression, substituting 4 for *a* and 8 for *b*.

$4^3 \div (2 \times 8)$

Step 2 Evaluate the expression.

$4^3 \div (2 \times 8)$ ← Perform operations in parentheses first.

$4^3 \div 16$ ← Then evaluate exponents.

$64 \div 16$ ← Then divide.

4

Solution **When $a = 4$ and $b = 8$, $a^3 \div (2 \times b) = 4$.**

Example 5

Each side of a cube is 5 units long. Find the surface area of the cube using the formula $A = 6s^2$, where s is the side length of the cube.

Strategy **Substitute the given side length of the cube in the formula. Then evaluate.**

Step 1 Rewrite the formula, substituting 5 for s.

$$A = 6(5^2)$$

Step 2 Evaluate the expression to the right of the equal sign.

$$A = 6(25)$$

$$A = 150$$

Solution **The surface area of the cube is 150 square units.**

Coached Example

The volume of a cube can be found with the formula $V = s^3$, where s is the side length of the cube. Tamara has two cubes. The first cube has a side length of 4 centimeters. The second has a side length of 7 centimeters. What are the volumes of Tamara's cubes?

Interpret the exponent in the formula.

s^3 means to use s as a factor ______ times.

$V = s^3 = s \times$ ______ $\times$ ______

Find the volume of the first cube.

Substitute ______ for s.

$V =$ ______ $\times$ ______ $\times$ ______ $=$ ______

Find the volume of the second cube.

Substitute ______ for s.

$V =$ ______ $\times$ ______ $\times$ ______ $=$ ______

Tamara's first cube has a volume of ______ cubic centimeters, and her second cube has a volume of ______ cubic centimeters.

Lesson Practice

Choose the correct answer.

1. What is the value of the expression below?

$20 + 8 - 4^2$

A. 12
B. 24
C. 28
D. 44

2. What is the value of the expression below?

$6 \div 2 - 1$

A. 6
B. 5
C. 3
D. 2

3. What is the value of the expression below when $a = 2$ and $b = 4$?

$3a + b$

A. 9
B. 10
C. 18
D. 24

4. What is the value of the expression below when $k = 4$?

$18 - k^2$

A. 2
B. 8
C. 16
D. 128

5. What is the value of the expression below when $m = 9$ and $n = 3$?

$m^2 \div (n + 6)$

A. 84
B. 33
C. 15
D. 9

6. What is the value of the expression below when $x = 6$ and $y = 2$?

$xy - y^3$

A. 4
B. 15
C. 54
D. 1,000

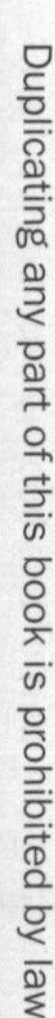

7. What is the area of a square with a side length of 11 inches? Use the formula $A = s^2$, where s is the side length of the square.

A. 22 square inches

B. 44 square inches

C. 121 square inches

D. 1,331 square inches

8. What is the volume of a cube with a side length of 17 centimeters? Use the formula $V = s^3$, where s is the side length of the cube.

A. 20 cubic centimeters

B. 51 cubic centimeters

C. 289 cubic centimeters

D. 4,913 cubic centimeters

9. Use the expression $(8g - 4h) \div h^2$ to answer the questions below.

A. What is the value of the expression when $g = 6$, and $h = 3$?

____

B. Explain how you used the order of operations to find the value of the expression.

____

____

____

Common Core State Standards: 6.EE.3, 6.EE.4

Work with Expressions

Getting the Idea

Since variables represent numbers, number properties apply to variables as well. For example, the commutative property allows you to say both $8 + 2 = 2 + 8$ and $a + b = b + a$.

You can use number properties to write and identify **equivalent expressions**. Expressions are equivalent if they name the same number regardless of which value a variable stands for. Remember that if a variable appears more than once in an expression, that variable refers to the same number in each instance.

Example 1

Write an equivalent expression for $2(5 + n)$.

Strategy **Use the distributive property of multiplication over addition.**

Step 1 Write the expression as the sum of two sets of factors.

$2(5 + n) = (2 \times 5) + (2 \times n)$

Step 2 Multiply each set of factors.

$(2 \times 5) + (2 \times n)$

$10 + 2n$

Solution **An equivalent expression for $2(5 + n) = 10 + 2n$.**

Example 2

Write an equivalent expression for $56x - 63$.

Strategy **Use the distributive property of multiplication over subtraction.**

Step 1 Identify the greatest common factor (GCF) of the terms in the expression.

Factors of 56: 1, 2, 4, 7, 8, 14, 28, 56

Factors of 63: 1, 3, 7, 9, 21, 63

The GCF of 56 and 63 is 7.

Step 2 Use the GCF and the distributive property to write an equivalent expression.

$56x - 63 = 7(8x) - 7(9) = 7(8x - 9)$

Solution **An equivalent expression for $56x - 63$ is $7(8x - 9)$.**

Example 3

Write an equivalent expression for $a + a + a$.

Strategy **Use number properties to write an equivalent expression.**

Step 1 What is the coefficient of each a term?

Remember that the identity property of multiplication states that the product of any number and 1 is that number.

So, $a = 1a$.

The coefficient of each a term is 1.

Step 2 Use the distributive property of multiplication over addition.

$a + a + a = a \times (1 + 1 + 1)$

Step 3 Simplify the expression in parentheses.

$a \times (1 + 1 + 1) = a \times 3$

Step 4 Use the commutative property of multiplication.

$a \times 3 = 3 \times a$

$3 \times a = 3a$

Solution **An equivalent expression for $a + a + a$ is $3a$.**

Like terms are terms that have the same variable raised to the same power. For example, in the expression $x^2 + 3x + 2x + 4x^2$, x^2 and $4x^2$ are like terms, and $3x$ and $2x$ are like terms. You can simplify an expression by combining like terms.

Example 4

Simplify $10x + 6y + 4x$.

Strategy **Simplify using like terms.**

Step 1 Identify the like terms in the expression.

The like terms are $10x$ and $4x$.

Step 2 Use the commutative property of addition to rewrite the expression.

$10x + 6y + 4x = 10x + 4x + 6y$

Step 3 Use the associative property of addition to group like terms.

$(10x + 4x) + 6y$

Step 4 Use the distributive property of multiplication over addition.

$(10x + 4x) + 6y = x(10 + 4) + 6y = x(14) + 6y$

Step 5 Use the commutative property of multiplication.

$x(14) + 6y = 14x + 6y$

Solution **$10x + 6y + 4x$ can be simplified as $14x + 6y$.**

Coached Example

Are the expressions $9n - 27$ and $3(3n - 9)$ equivalent?

To simplify $3(3n - 9)$, use the ____________________ property of multiplication over ____________________.

$3(3n - 9) = 3(\_\_\_) - \_\_\_(\_\_\_) = \_\_\_ - \_\_\_$

Are the expressions equivalent? ____________

Check your work by substituting different values for n into each expression.

Try $n = 4$ in both expressions first.

$9n - 27 = 9(\_\_\_) - 27 = \_\_\_ - \_\_\_ = \_\_\_$

$3(3n - 9) = 3(3 \times \_\_\_) - 3(9) = 3(\_\_\_) - \_\_\_ = \_\_\_ - \_\_\_ = \_\_\_$

Now try $n = 10$ in both expressions.

$9n - 27 = 9(\_\_\_) - 27 = \_\_\_ - \_\_\_ = \_\_\_$

$3(3n - 9) = 3(3 \times \_\_\_) - 3(9) = 3(\_\_\_) - \_\_\_ = \_\_\_ - \_\_\_ = \_\_\_$

Did both expressions have the same value for each given value of n? ____________

The expressions $9n - 27$ and $3(3n - 9)$ ____________ equivalent.

Lesson Practice

Choose the correct answer.

1. Which expression is equivalent to $b + b + b + b$?

A. $4b$

B. $b + 4$

C. b^4

D. $b \div 4$

2. Which expression is equivalent to $7(3 + g)$?

A. $21 + g$

B. $10g$

C. $21 + 7g$

D. $21g$

3. Which expression is **not** equivalent to $5x + 6$?

A. $4x + 7 + x - 1$

B. $3x + 3 + 2x + 3$

C. $5(x + 1) + 1$

D. $x(5 + 6)$

4. Which expression is equivalent to $4t + 3t$?

A. $7t^2$

B. $7 + 2t$

C. $7t$

D. $12t$

5. Which expression is equivalent to $9c + 12d + 2c$?

A. $18c^2 + 12d$

B. $11c + 12d$

C. $11c^2 + 12d$

D. $23cd$

6. Which expression is **not** equivalent to $4k + 12$?

A. $3k + 4 + k + 8$

B. $3(k + 3) + 3$

C. $4(k + 3)$

D. $2(2k + 5) + 2$

7. Which expression is equivalent to $6(p + 5)$?

   A. $30 + 6p$

   B. $30p$

   C. $30 + p$

   D. $11p$

8. For which value or values are the expressions $15k + 9$ and $3(2k + 3) + 9k$ equivalent?

   A. no values

   B. 3

   C. 3, 5, and 8

   D. all values

9. The lengths of the sides of a triangle are represented by $3m$, $3m$, and $3m$.

   A. What is an expression, in simplest form, for the perimeter of the triangle?

   ____

   B. Use the distributive property to write an equivalent expression for the perimeter of the triangle.

   ____

   ____

   ____

Common Core State Standards: 6.EE.5, 6.EE.7

Equations

Getting the Idea

An **equation** is a mathematical statement that says two expressions are equal. An equation has an equal sign (=).

A variable can be used to represent an unknown number in an equation. Here are some examples of equations with variables.

$6z = 36$ $\quad n = \frac{3}{4}p$ $\quad a + 3 = 11$ $\quad d - 5 = 22$ $\quad \frac{c}{7} = 8$

Equations can be used to represent real-world situations. For example, imagine that Holly used 90 chocolate chips to make 15 cookies, and each cookie contained the same number of chips. You could use the equation $15x = 90$ to represent this situation, with x representing the unknown number of chocolate chips in each cookie.

Example 1

Jerry scored a total of 12 points in a basketball game. He scored the same number of points each time he made a basket. The equation $4x = 12$ represents this situation. Find the value of x, the number of points Jerry scored with each basket.

Strategy **Use substitution to find the value of *x*.**

Step 1 Choose a set of possible values for x.

Jerry could have scored 1, 2, or 3 points with each basket.

Step 2 Substitute the possible values for x. Try 1 first.

$4x = 12$

$4(1) \stackrel{?}{=} 12$

$4 \neq 12$

Step 3 Substitute 2 for x.

$4x = 12$

$4(2) \stackrel{?}{=} 12$

$8 \neq 12$

Step 4 Substitute 3 for x.

$4(3) \stackrel{?}{=} 12$

$12 = 12$ ✓

If $4x = 12$, then $x = 3$.

Solution **Jerry scored 3 points with each basket.**

Example 2

What is the value of x in the equation $x + 27 = 75$?

Strategy **Use substitution to solve.**

Step 1 Estimate a solution.

$50 + 25 = 75$, so x will be less than, but close to, 50.

Step 2 Try the following values for x: 47, 48, and 49.

Substitute 47 for x.	Substitute 48 for x.	Substitute 49 for x.
$x + 27 = 75$	$x + 27 = 75$	$x + 27 = 75$
$47 + 27 \stackrel{?}{=} 75$	$48 + 27 \stackrel{?}{=} 75$	$49 + 27 \stackrel{?}{=} 75$
$74 \neq 75$	$75 = 75$ ✓	$76 \neq 75$

Solution $\boldsymbol{x = 48}$

To solve an equation, you need to isolate the variable. You can use **inverse operations** to isolate the variable. Inverse operations are related operations that undo each other, such as addition and subtraction, or multiplication and division.

Example 3

Solve for d.

$\frac{d}{4} = 16$

Strategy **Use an inverse operation to isolate the variable.**

Step 1 Identify the inverse operation.

$\frac{d}{4}$ is equivalent to $d \div 4$.

Multiplication is the inverse operation of division.

Step 2 Multiply both sides of the equation by 4.

$$\frac{d}{4} = 16$$

$$\frac{d}{4} \times 4 = 16 \times 4$$

$$d = 64$$

Solution $\boldsymbol{d = 64}$

Example 4

What value of y makes this equation true?

$y - 19 = 24$

Strategy **Use an inverse operation to isolate the variable.**

Step 1 Identify the inverse operation.

Addition is the inverse operation of subtraction.

Step 2 Add 19 to both sides of the equation.

$$y - 19 = 24$$
$$y - 19 + 19 = 24 + 19$$
$$y = 43$$

Solution $\mathbf{y = 43}$

Some equations may contain fractions. Remember that dividing by a fraction is the same as multiplying by its reciprocal.

Example 5

What is the value of z in the equation $\frac{1}{3}z = 26$?

Strategy **Use inverse operations to isolate the variable.**

Step 1 Divide both sides by $\frac{1}{3}$ to isolate the variable.

Multiply by the reciprocal of $\frac{1}{3}$.

The reciprocal of $\frac{1}{3}$ is $\frac{3}{1}$, or 3.

$$\frac{1}{3}z \times 3 = 26 \times 3$$
$$z = 78$$

Step 2 Check your answer.

Substitute 78 for z in the original equation.

$$\frac{1}{3}(78) = 26$$
$$26 = 26 \checkmark$$

Solution $\mathbf{z = 78}$

Coached Example

Gavin worked 16 hours last week and earned \$192. The equation $16d = 192$ can be used to find d, the number of dollars he earns per hour. What is Gavin's hourly wage? Use inverse operations to solve for d.

What operation is used in the term $16d$? ____________

The inverse of multiplication is ________________.

To isolate the variable, ____________ both sides of the equation by ________.

Solve for d.

$16d = 192$

$16d \div$ ________ $=$ ________ $\div$ ________

$d =$ ________

Gavin's hourly wage is \$________.

Lesson Practice

Choose the correct answer.

1. Which is the solution to $2f = 32$?

A. 12

B. 14

C. 16

D. 18

2. Which step should be taken to isolate the variable in the following equation?

$7d = 49$

A. Add 7 to both sides of the equation.

B. Subtract 7 from both sides of the equation.

C. Multiply both sides of the equation by 7.

D. Divide both sides of the equation by 7.

3. What is the value of c in the following equation?

$29 + c = 62$

A. 33

B. 43

C. 81

D. 91

4. What is the value of j in the following equation?

$j - 87 = 165$

A. 78

B. 88

C. 242

D. 252

5. What is the value of n in the following equation?

$22n = 418$

A. 12

B. 14

C. 19

D. 24

6. What is the value of k in the following equation?

$\frac{1}{5}k = 5$

A. 0

B. 1

C. 10

D. 25

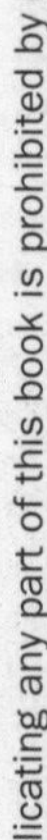

7. What is the value of a in the following equation?

$3a = 15$

A. 1

B. 5

C. 6

D. 45

8. A music teacher bought 19 recorders. She spent a total of \$57. Each recorder was the same price. The equation $19r = 57$ can be used to find r, the price of each recorder in dollars. What was the price of each recorder?

A. \$3

B. \$4

C. \$38

D. \$76

9. Trista solved an equation for x. Her solution is shown below.

$$36 + x = 54$$

$$36 + x - 36 = 54 + 36$$

$$x = 90$$

A. Trista's solution is incorrect. What is the correct value of x? Show your work.

____

B. What error did Trista make?

____

____

____

Common Core State Standard:
6.EE.9

Dependent and Independent Variables

Getting the Idea

Two variables are often related in real-world situations. For example, sales of ice cream cones may be related to the temperature. There are different ways to show how two variables are related.

The equation $y = x + 1$ shows how the variables x and y are related. In the equation $y = x + 1$, x is called the **independent variable** and y is called the **dependent variable**, since the value of y depends on the value of x.

You also can use a table or a graph to show how two variables are related.

Example 1

Create a table of values for the equation $y = x + 1$. Include at least four ordered pairs in the table.

Strategy **Make a table of values.**

Step 1 Choose several values for x.

Choose whole numbers, such as 2, 4, 6, and 8.

Step 2 Create a table of values.

Substitute each x-value into the equation to find its corresponding y-value.

x	$y = x + 1$	y	(x, y)
2	$y = 2 + 1 = 3$	3	(2, 3)
4	$y = 4 + 1 = 5$	5	(4, 5)
6	$y = 6 + 1 = 7$	7	(6, 7)
8	$y = 8 + 1 = 9$	9	(8, 9)

The last column of the table shows ordered pairs of x- and y-values for the equation.

Solution **The table of values in Step 2 above shows four ordered pairs for $y = x + 1$.**

Example 2

Use the ordered pairs you found in Example 1 to graph $y = x + 1$.

Strategy **Plot a point for each ordered pair, then connect them.**

Step 1 Plot points for each ordered pair on a coordinate grid.

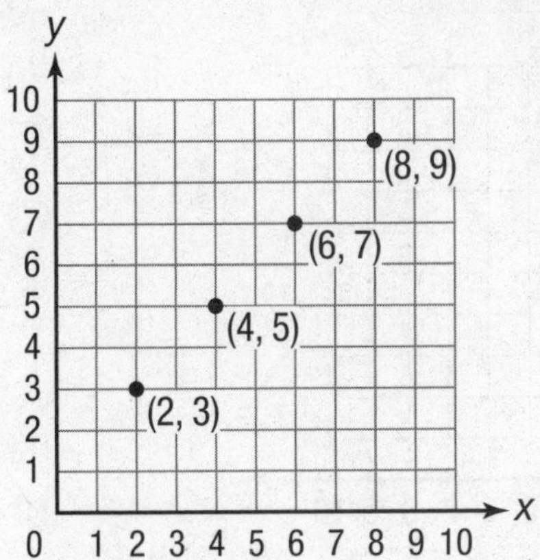

Step 2 Connect the points.

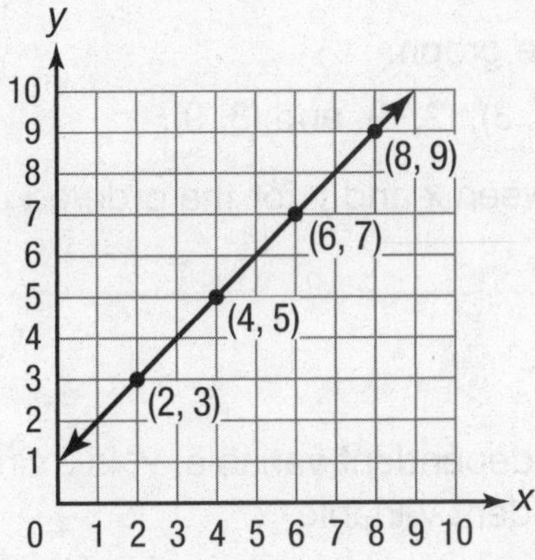

Solution **The graph of $y = x + 1$ is shown in Step 2 above.**

You can write an equation for the graph of a line.

Example 3

Write an equation to represent the graph of the line below.

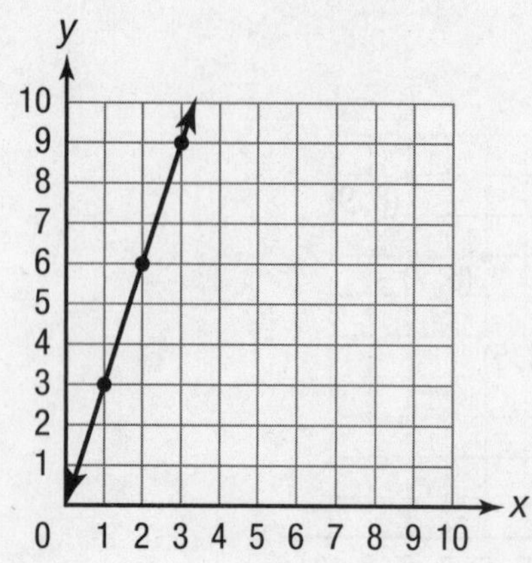

Strategy **Identify the ordered pairs on the graph. Determine how the variables are related. Then write an equation.**

Step 1 Identify the ordered pairs on the graph.

The graph goes through (1, 3), (2, 6), and (3, 9).

Step 2 Determine the relationship between x and y for the ordered pairs.

For (1, 3): $1 \cdot \mathbf{3} = 3$

For (2, 6): $2 \cdot \mathbf{3} = 6$

For (3, 9): $3 \cdot \mathbf{3} = 9$

For each ordered pair, the dependent variable, y, is 3 times the corresponding independent variable, x.

Step 3 Write an equation.

The graph of the line can be represented as $y = 3 \cdot x$ or $y = 3x$.

Solution **The equation $y = 3x$ represents the graph of the line.**

Coached Example

Create a table of values to represent $y = x + 2$. Use it to find four ordered pairs that represent the equation. Then graph the equation.

Choose integers such as 1, 3, 5, and 7.

Then substitute each *x*-value into the equation to find its corresponding *y*-value.

x	$y = x + 2$	y	(x, y)
1	$y = 1 + 2 =$ ____	____	(1, ____)
3	$y =$ ____ $+ 2 =$ ____	____	(3, ____)
5	$y =$ ____ $+ 2 =$ ____	____	(5, ____)
7	$y =$ ____ $+ 2 =$ ____	____	(7, ____)

Now plot the ordered pairs on the grid below. Draw a line to connect the points and graph the equation.

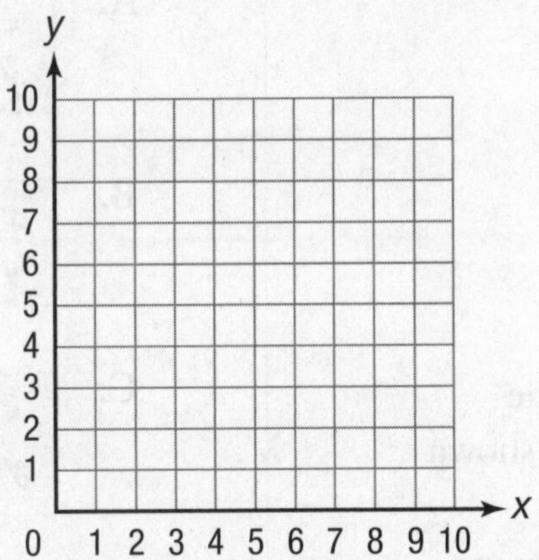

Four ordered pairs for $y = x + 2$ are (1, ___), (3, ___), (5, ___), and (7, ___).

The graph of $y = x + 2$ is shown above.

Lesson Practice

Choose the correct answer.

1. Which equation best represents the relationship between x and y shown in the table?

x	0	1	2	3
y	1	5	9	13

A. $y = x + 1$
B. $y = x \cdot 5$
C. $y = 4x + 1$
D. $y = 5x + 1$

2. Which ordered pair does **not** represent the equation $y = x + 3$?

A. (1, 4)
B. (3, 9)
C. (12, 15)
D. (18, 21)

3. Which equation represents the relationship between x and y shown in the table?

x	6	12	18	24
y	2	4	6	8

A. $y = x - 4$
B. $y = 3x$
C. $y = \frac{1}{3}x$
D. $y = \frac{1}{6}x$

4. Which table below best represents the relationship between x and y shown in the graph?

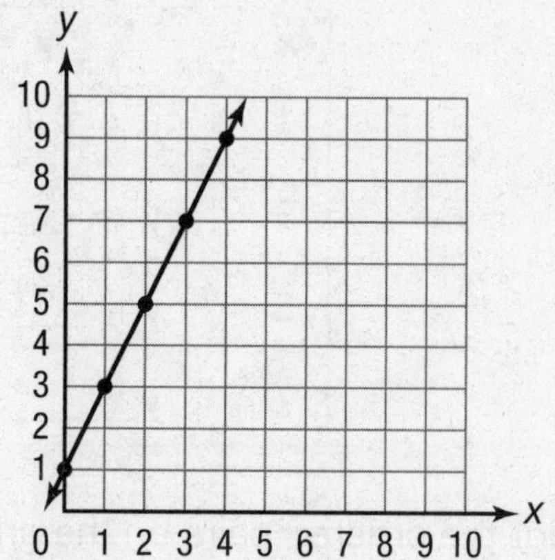

A.

x	0	1	2	3	4
y	1	3	5	7	9

B.

x	0	3	5	3	9
y	1	1	2	7	4

C.

x	1	3	5	7	9
y	0	1	2	3	4

D.

x	0	1	2	3	4
y	0	1	3	5	7

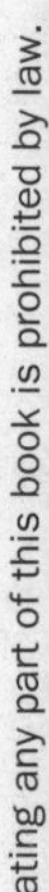

5. Use the equation $y = x - 1$ to answer the questions below.

A. Complete the table of values below for $y = x - 1$. Use the space provided in the second column to show your work.

x	$y = x - 1$	y	(x, y)
3			
5			
7			
9			

B. Graph $y = x - 1$ on the coordinate grid below. Explain in words how you graphed the equation.

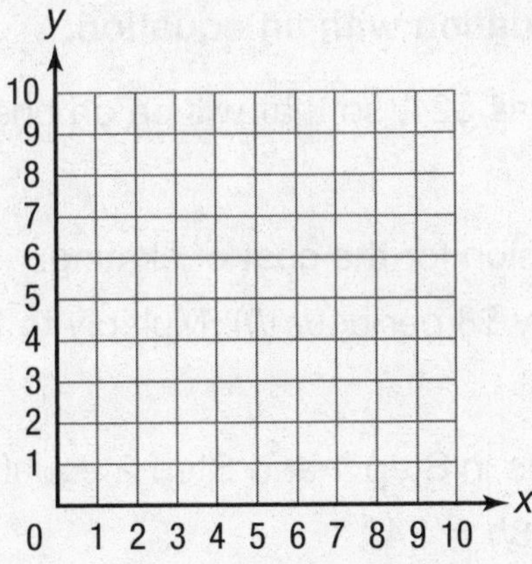

____

____

____

Common Core State Standards: 6.EE.6, 6.EE.7, 6.EE.9

Use Equations to Solve Problems

Getting the Idea

You can write an equation to help you solve a word problem by choosing a variable to represent the unknown quantity.

Example 1

It cost \$8 per hour to ice skate. Katy spent \$24 skating. Write an equation you can use to find h, the number of hours Katy skated. Then solve the equation.

Strategy **Express the situation with an equation.**

Step 1 The total cost was \$24, so that will be on one side of the equation.

24

Step 2 Write an expression for the cost of skating.

It costs Katy \$8 per hour (h). Multiply to find the cost.

$8 \times h$ or $8h$

Step 3 Set the quantities in Step 1 and Step 2 equal to each other.

$24 = 8h$ or $8h = 24$

Step 4 Solve the equation for h.

Divide both sides of the equation by 8.

$\frac{8h}{8} = \frac{24}{8}$

$h = 3$

Solution **The equation $8h = 24$ can be used to find the number of hours Katy skated. Katy skated for 3 hours.**

Example 2

Ted has 60 books on 6 shelves. The bottom shelf has 15 books. The other shelves have an equal number of books. Write an equation that can be used to find b, the number of books on each of the top 5 shelves. Then solve the equation.

Strategy **Express the situation with an equation.**

Step 1 There are a total of 60 books, so that will be on one side of the equation.

60

Step 2 Write an expression for the other side of the equation.

One shelf has 15 books. The other 5 shelves each have b books.

The expression $5b + 15$ represents the total number of books.

Step 3 Write the equation.

$60 = 5b + 15$, or $5b + 15 = 60$

Step 4 Solve the equation for b.

$5b + 15 - 15 = 60 - 15$ ← Subtract 15 from both sides of the equation.

$5b = 45$ ← Divide both sides of the equation by 5.

$\frac{5b}{5} = \frac{45}{5}$

$b = 9$

Solution **The equation $5b + 15 = 60$ can be used to find the number of books on each of the top 5 shelves. There are 9 books on each of the top 5 shelves.**

Example 3

A babysitter charges $5 per hour plus a flat fee of $10 for each babysitting job. How many hours did the babysitter work if she earned $35?

Strategy **Write an equation to represent the situation. Then solve the equation.**

Step 1 The amount received was $35, so that will be on one side of the equation.

35

Step 2 Write an expression for the number of hours and the flat fee.

The babysitter charges $5 per hour ($h$). The flat fee is $10.

The expression $5h + 10$ represents the total amount the babysitter earns on a job.

Step 3 Set the quantities in Step 1 and Step 2 equal to each other.

$35 = 5h + 10$ or $5h + 10 = 35$

Step 4 Solve the equation for h.

$5h + 10 - 10 = 35 - 10$ ← Subtract 10 from both sides of the equation.

$5h = 25$ ← Divide both sides of the equation by 5.

$\frac{5h}{5} = \frac{25}{5}$

$h = 5$

Solution **The babysitter worked for 5 hours.**

Example 4

The Davis family drove to their cousins' house for Thanksgiving weekend. Each hour, Mrs. Davis recorded how far they had driven from home. The table below shows her data.

Miles Driven

Number of Hours	Number of Miles
1	60
2	120
3	180
4	240
5	300

Graph the data in the table. Then write an equation to show how distance driven and time are related for the Davis family.

Strategy **Determine the dependent and independent variables. Graph the data. Write an equation.**

Step 1 Find the dependent and independent variables.

The distance that the Davis family has traveled depends on how many hours have passed since they left home.

Distance, or number of miles, is the dependent variable.

Time, or number of hours, is the independent variable.

Step 2 Graph the ordered pairs on a coordinate grid.

Use the y-axis to represent the dependent variable, number of miles.

Use the x-axis to represent the independent variable, number of hours.

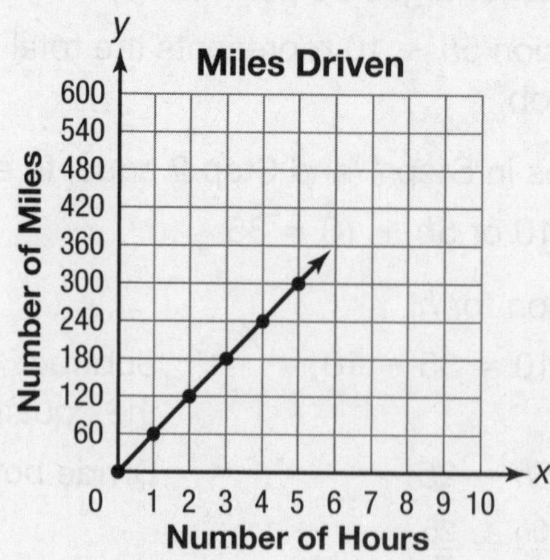

Step 3 Analyze the ordered pairs in the table and the graph.

How are the x- and y-values in each ordered pair related?

For (1, 60): $1 \cdot \mathbf{60} = 60$

For (2, 120): $2 \cdot \mathbf{60} = 120$

For (3, 180): $3 \cdot \mathbf{60} = 180$

For (4, 240): $4 \cdot \mathbf{60} = 240$

For (5, 300): $5 \cdot \mathbf{60} = 300$

Each y-value is 60 times the corresponding x-value.

Step 4 Write an equation.

$y = 60 \times x$ or $y = 60x$

Step 5 Interpret the equation.

The equation $y = 60x$ means that, for every hour, the Davis family traveled 60 miles. You could also say that the family was traveling at an average speed of 60 miles per hour.

Solution **The equation $y = 60x$ represents the data in the table and is graphed in Step 2. The Davis family was traveling at an average speed of 60 miles per hour.**

Coached Example

Mr. Alvarez bought concert tickets for his family of 6. The total cost of the tickets was $320. There was a $20 handling fee. Write an equation that can be used to find *t*, the cost of each ticket. Then solve the equation.

How many tickets were purchased? ______

What variable represents the cost of each ticket? ______

Write an expression that represents the cost of 6 tickets. ______

The handling fee of $20 must be ______________ to the cost of the tickets.

Write an expression that represents the total cost of the tickets. ______ + ______

What was the total cost of the tickets, in dollars? ______

Now write the equation: __________ + __________ = __________

Solve for *t*.

First subtract __________ from both sides of the equation.

$6t + 20 -$ __________ $= 320 -$ __________

The new equation is: __________ = __________

__________ both sides of the equation by __________.

$\frac{6t}{\_\_\_} =$ ______

$t =$ ______

The equation ______________ can be used to find the cost of each ticket.

Each ticket costs $________.

Lesson Practice

Choose the correct answer.

1. After Kwan spent \$5 dollars of the money he earned for mowing lawns, he had \$15 left. Let m equal the amount Kwan earned mowing lawns. Which shows the equation that represents the situation and the amount Kwan earned mowing lawns?

   A. $m - 5 = 15$; \$20

   B. $m + 5 = 15$; \$20

   C. $5m = 15$; \$3

   D. $\frac{m}{5} = 15$; \$75

2. Pete has 4 times as many model cars as Steve. Pete has 24 model cars. Let s equal the number of model cars Steve has. Which shows the equation that represents the situation and the number of model cars Steve has?

   A. $\frac{s}{4} = 24$; 96 model cars

   B. $24s = 4$; $\frac{1}{6}$ model car

   C. $4s = 24$; 6 model cars

   D. $s + 4 = 24$; 20 model cars

3. Kristin worked a total of 10 hours over two days. She worked 6 hours the first day and h hours the second day. Which shows the equation that represents the situation and the number of hours she worked the second day?

   A. $6h = 10$; 1.5 hours

   B. $\frac{h}{10} = 6$; 60 hours

   C. $h - 6 = 10$; 16 hours

   D. $6 + h = 10$; 4 hours

4. Patel has 5 less than 4 times as many trophies as Horatio. He has 19 trophies in all. How many trophies does Horatio have?

   A. 3

   B. 6

   C. 71

   D. 91

5. A restaurant is offering a $10-off special per table if three or more dinners are ordered. Four friends each ordered the same dinner and spent a total of $150. What was the cost of each dinner?

A. $13
B. $30
C. $35
D. $40

6. The Downtown Theater has 413 seats, which is 8 more than 3 times as many seats as the Uptown Theater has. How many seats are there in the Uptown Theater?

A. 135
B. 405
C. 1,247
D. 3,301

7. A boat rental company charges $25 per hour. A life jacket can be rented for $10 for the day. Larry spent $85 in all including the cost of a life jacket rental. How many hours did Larry rent the boat?

A. 5 hours
B. 4 hours
C. $3\frac{1}{2}$ hours
D. 3 hours

8. Tristan has 20 less than 3 times as many DVDs as Anna. If Tristan has 55 DVDs, how many DVDs does Anna have?

A. 25
B. 32
C. 38
D. 145

9. Marlene studied 10 minutes more than 4 times as long as Brianna.

A. What expression can you write to represent the situation? Explain what your variable represents.

____

____

B. How many minutes did Brianna study if Marlene studied for 150 minutes? Show and explain how you found your answer.

____

____

____

Common Core State Standards: 6.EE.5, 6.EE.6, 6.EE.8

Inequalities

Getting the Idea

An **inequality** is a mathematical statement that compares two expressions and includes an inequality symbol. The different inequality symbols are shown below.

$>$ means **"is greater than."** $\geq$ means **"is greater than or equal to."**

$<$ means **"is less than."** $\leq$ means **"is less than or equal to."**

Example 1

Write an inequality to represent this phrase:

a number less than 5

Identify the variable you used and explain what the variable represents.

Strategy **Translate the words into an inequality.**

Step 1 Translate the words into an inequality.

Let the variable x represent the unknown number.

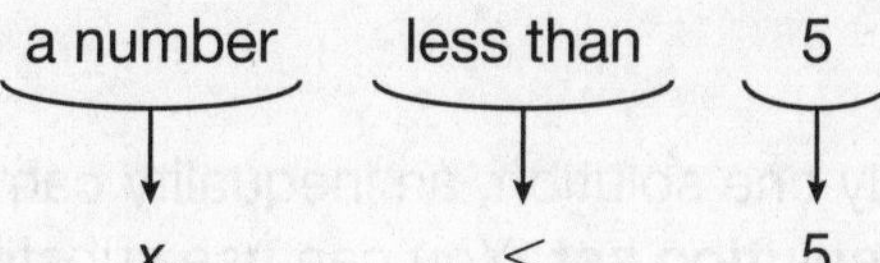

Step 2 Explain what the variable, x, represents.

The variable x represents the set of all numbers that are less than 5.

So, 4, 0, and -10 are all possible values of x, because all those numbers are less than 5.

Solution **The phrase "a number less than 5" can be represented as $x < 5$. The variable, x, represents the set of all numbers that are less than 5.**

Sometimes, you may need to translate a real-world problem into an inequality. The table below shows some key words that may help you decide which inequality symbol to use.

$>$	$<$	$\geq$	$\leq$
greater than more than	less than fewer than	at least no less than	at most no more than

Example 2

A pastry chef wants to spend no more than \$100 on almonds. Almonds cost \$4 per pound. Write an inequality to represent p, the number of pounds of almonds the chef could buy.

Strategy **Translate the situation into an inequality.**

Step 1 Identify and interpret the key words.

The variable p represents the number of pounds of almonds the chef could buy.

Almonds cost \$4 per pound, so the total cost of p pounds of almonds is $4p$.

The key words "no more than" indicate using the less than or equal symbol ($\leq$).

Step 2 Write an inequality.

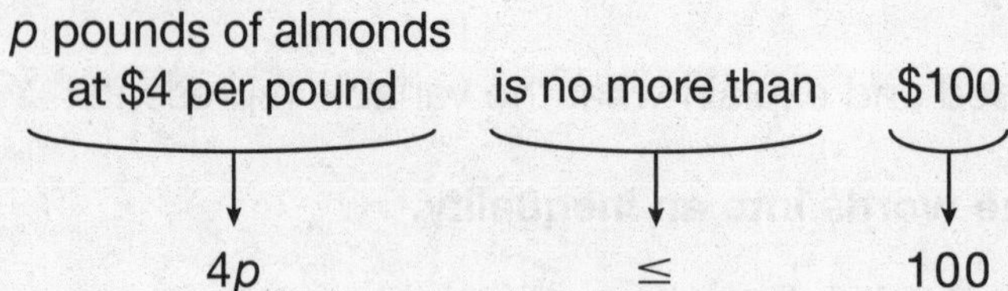

Solution **The inequality $4p \leq 100$ represents the number of pounds of almonds the chef could buy.**

Unlike an equation, which usually has only one solution, an inequality can have many solutions. These solutions are called the **solution set**. You can use substitution to determine if a number is in the solution set or not.

Example 3

Which of the following are possible solutions for the inequality below: 10.5, 25, or 30?

$4p \leq 100$

Strategy **Substitute each value for p in the inequality. Determine which value or values make the inequality true.**

Step 1 Substitute 10.5 for p.

$$4p \leq 100$$
$$4(10.5) \overset{?}{\leq} 100$$
$$42 \leq 100 \checkmark$$

10.5 is a solution for the inequality.

Step 2 Substitute 25 for p.

$4p \leq 100$

$4(25) \overset{?}{\leq} 100$

$100 \leq 100$ ✓

Since the inequality uses the $\leq$ symbol, any number less than or equal to 100 is a solution. So 25 is a solution to the inequality.

Step 3 Substitute 30 for p.

$4p \leq 100$

$4(30) \overset{?}{\leq} 100$

$120 \leq 100$ is false, because $120 > 100$.

So, 30 is not a solution to the inequality.

Solution **The numbers 10.5 and 25 are solutions for the inequality $4p \leq 100$. The number 30 is not a solution.**

To solve an inequality, follow the same steps as you would to solve an equation. Remember, whatever you do to one side of the inequality must also be done to the other side.

Note: If you multiply or divide both sides of an inequality by a negative number, you need to reverse the inequality symbol.

Example 4

Solve for p: $4p \leq 100$

Strategy **Use inverse operations to isolate the variable.**

Since p is multiplied by 4, divide both sides by 4.

$4p \leq 100$

$\frac{4p}{4} \leq \frac{100}{4}$

$p \leq 25$

Solution **The solution set for the inequality is all numbers less than or equal to 25, or $p \leq 25$.**

Example 5

What is the solution set for $\frac{n}{3} \geq 12$?

Strategy **Use inverse operations to isolate the variable.**

Step 1 Since n is divided by 3, multiply both sides by 3.

$$\frac{n}{3} \geq 12$$

$$\frac{n}{3} \times 3 \geq 12 \times 3$$

$$n \geq 36$$

Step 2 Use substitution to check your answer.

Substitute a number that is less than 36, such as 33.

$$\frac{33}{3} \overset{?}{\geq} 12$$

$11 \geq 12$ is false, because $11 < 12$.

Now substitute 36.

$$\frac{36}{3} \overset{?}{\geq} 12$$

$12 \geq 12$ ✓

Substitute a number that is greater than 36, such as 39.

$$\frac{39}{3} \overset{?}{\geq} 12$$

$13 \geq 12$ ✓

Only the numbers that were greater than or equal to 36 made the inequality true.

Solution **The solution set for the inequality is $n \geq 36$.**

You can graph the solution set of an inequality on a number line. Draw a circle and an arrow to show all the numbers that are part of the solution.

- An open circle indicates that the circled number is not a solution for the inequality. Use an open circle when the symbol is $>$ or $<$.

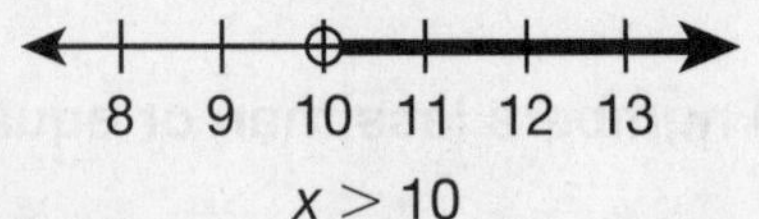

$x > 10$

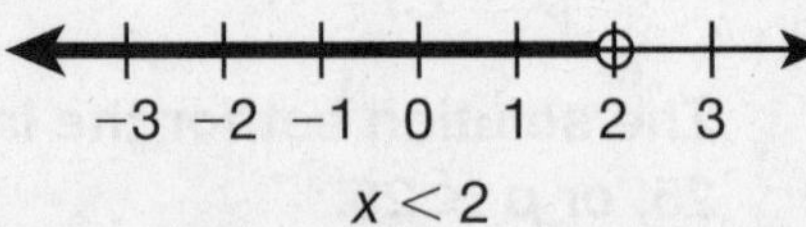

$x < 2$

- A closed circle indicates that the circled number is a solution. Use a closed circle when the symbol is $\geq$ or $\leq$.

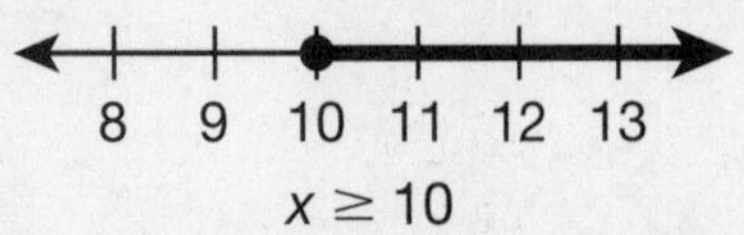

$x \geq 10$

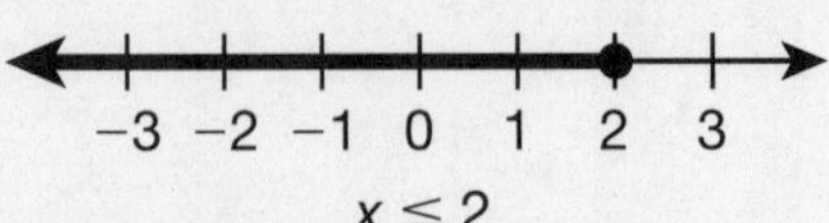

$x \leq 2$

Example 6

Solve and graph the inequality $b + 13 > 17$. Then identify 3 possible values for b.

Strategy **Use inverse operations to isolate the variable. Then graph the inequality.**

Step 1 Isolate the variable.

$$b + 13 > 17$$
$$b + 13 - 13 > 17 - 13$$
$$b > 4$$

Step 2 Draw a number line and make a circle at 4.

The symbol is $>$, so draw an open circle at 4 to show that 4 is not a solution.

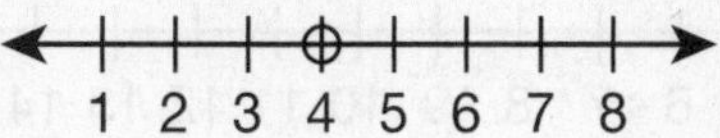

Step 3 Draw an arrow on the number line to show the solution set.

Since all values greater than 4 are solutions, draw an arrow to the right.

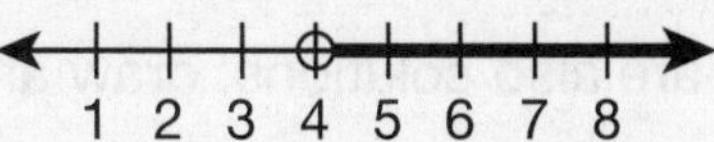

Step 4 Identify three solutions for the inequality.

5, $6\frac{1}{2}$, and 8 are all solutions because they are all part of the graph.

Solution **The solution set to $b + 13 > 17$ is $b > 4$, which is graphed in Step 3 above. Four possible values for b are 5, $6\frac{1}{2}$, and 8.**

Coached Example

Solve and graph the inequality $z - 8 \leq 5$.

Use inverse operations to isolate the variable.

Since _______ is subtracted from z, add _______ to both sides.

$$z - 8 \leq 5$$

$$z - 8 + \_\_\_\_\_\_\_ \leq 5 + \_\_\_\_\_\_\_$$

$$z \leq \_\_\_\_\_\_\_$$

Graph the solution set on the number line below.

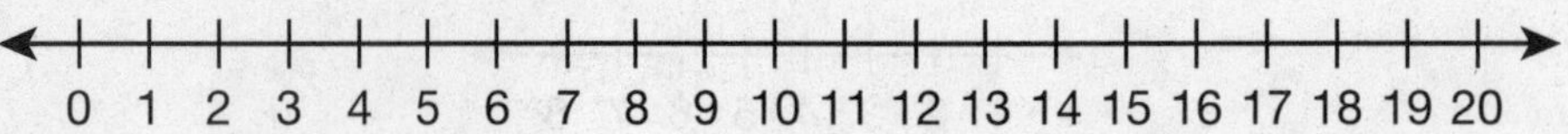

Since the symbol is $\leq$, draw a(n) _________ circle at _____.

This shows that _____ is a solution.

Since all values less than _____ are also solutions, draw an arrow to the ___________.

The solution set for the inequality is $z \leq$ _____. Its graph is shown above.

Lesson Practice

Choose the correct answer.

1. Which inequality best represents this phrase?

a number greater than -1

A. $x > -1$
B. $x < -1$
C. $x \geq -1$
D. $x \leq -1$

2. Which inequality best represents this phrase?

a number less than or equal to 0

A. $n = 0$
B. $n < 0$
C. $n \leq 0$
D. $n \geq 0$

3. Five friends had lunch together. Their total bill was x dollars, including tax and tip. They shared the cost equally and each friend paid less than \$10. Which inequality shows the possible solutions for x, the total amount of the bill?

A. $x > 50$
B. $x < 50$
C. $x \geq 50$
D. $x \leq 50$

4. A red block and a blue block are on a scale. The red block weighs 9 ounces. The total weight of both blocks is at most 16 ounces. Which inequality best represents b, the possible weight of the blue block in ounces?

A. $b \geq 7$
B. $b \leq 7$
C. $b \geq 25$
D. $b \leq 25$

5. Which number is **not** a solution for the inequality below?

$$\frac{x}{2} \geq 12$$

A. 6
B. 24
C. 25
D. 100

6. Which graph shows the solution set for this inequality?

$$r + 9 < 12$$

A. 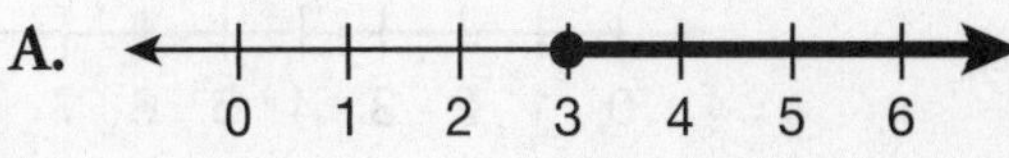

B. 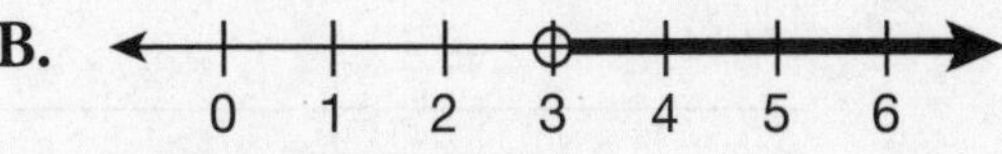

C. 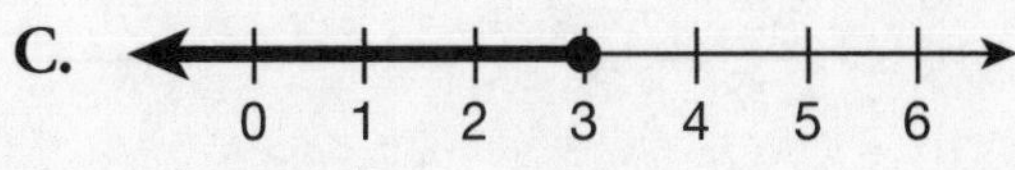

D. 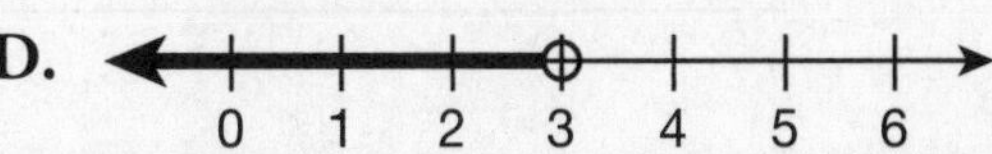

7. Which number is a solution for the inequality below?

$3n < 18$

A. 0
B. 6
C. 12
D. 18

8. Which graph shows the solution set for this inequality?

$8z \geq 16$

A. 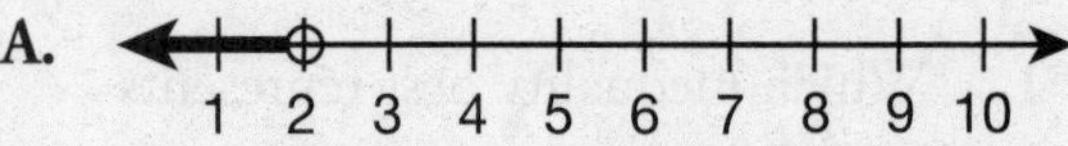

B. 

C. 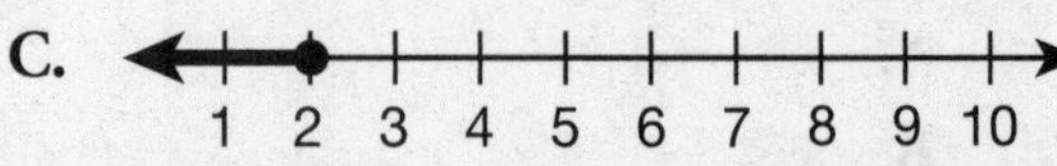

D. 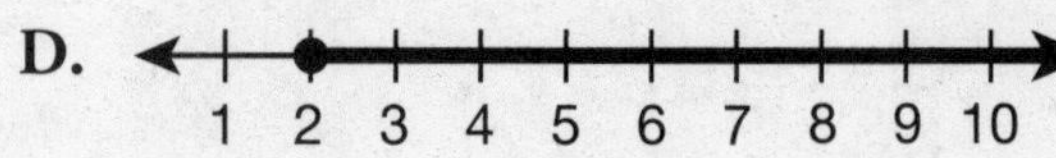

9. Mrs. Perry hires a landscaper that charges $15 per hour. The landscaper says that the total charge for the work Mrs. Perry wants done will be at least $120.

A. Write an inequality to show *h*, the number of hours it may take for the landscaper to do the work for Mrs. Perry. Solve the inequality and show your work below.

B. Graph the inequality on the number line below. Then use the graph to explain if it could take the landscaper 9.5 hours to complete the work for Mrs. Perry.

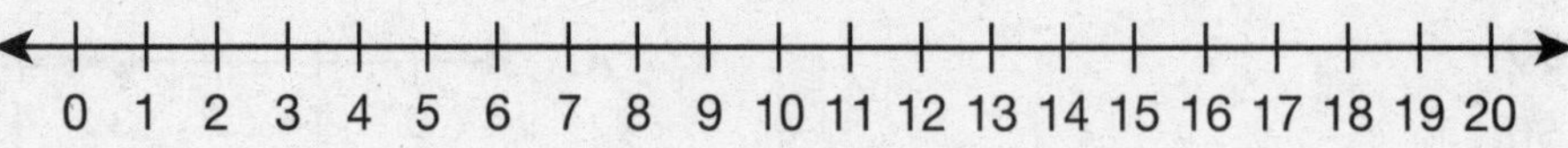

Domain 3: Cumulative Assessment for Lessons 17–23

1. Which expression represents "subtract a number n from 10, then multiply by 9 cubed"?

A. $9^3 \times (n - 10)$

B. $9^3 + 10n$

C. $9^3 - 10n$

D. $9^3 \times (10 - n)$

2. What is the value of the expression below when $b = 4$?

$7b + 6$

A. 34

B. 46

C. 53

D. 80

3. Which expression is equivalent to $5(9 + t)$?

A. $45t$

B. $45 + 5t$

C. $45 + t$

D. $14t$

4. Which number is a solution for the inequality below?

$\frac{x}{4} \geq 20$

A. 6

B. 16

C. 25

D. 90

5. Vikram needs to buy 72 juice boxes for a community picnic. At the store, juice boxes are sold in packs of 3 juice boxes each. Which shows the equation that represents the situation and the number of packs of juice boxes Vikram needs to buy?

A. $\frac{j}{3} = 72$; 216 juice packs

B. $j - 3 = 72$; 75 juice packs

C. $3j = 72$; 24 juice packs

D. $j + 3 = 72$; 72 juice packs

6. Which equation best represents the relationship between x and y shown in the table?

x	0	1	2	3
y	1	3	5	7

A. $y = x + 1$

B. $y = 3x$

C. $y = 3x - 2$

D. $y = 2x + 1$

7. Angelina wants to buy a pair of jeans and a sweater that costs \$38. She does not want to spend more than \$80 for the jeans and the sweater. Which inequality best represents j, the amount that Angelina can spend on the jeans?

A. $j > 42$

B. $j \geq 42$

C. $j < 42$

D. $j \leq 42$

8. What is the value of y in the following equation?

$\frac{1}{3}y = 9$

A. 3

B. 9

C. 27

D. 81

9. Describe the expression $12 + (n \div 6)$ in words.

10. The equation $y = x - 3$ describes how the variables x and y are related.

A. Complete the table of values below for $y = x - 3$. Show all your work.

x	$y = x - 3$	y	(x, y)
3			
5			
7			
9			

B. Graph $y = x - 3$ on the coordinate grid below.

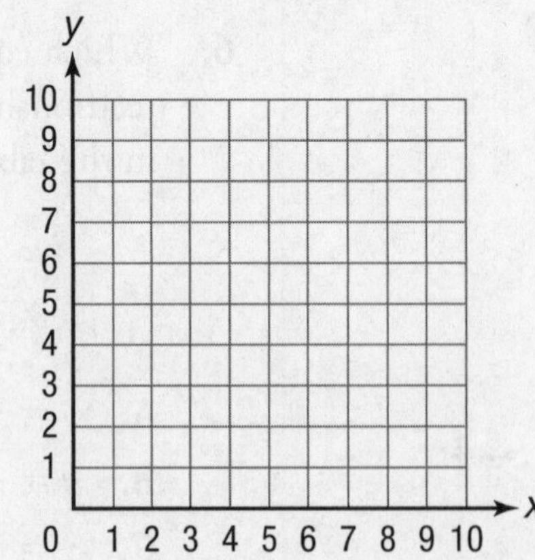

Domain 4

Geometry

Domain 4: Diagnostic Assessment for Lessons 24–31

Domain 4: Cumulative Assessment for Lessons 24–31

Domain 4: Diagnostic Assessment for Lessons 24–31

1. What is the area of the triangle below?

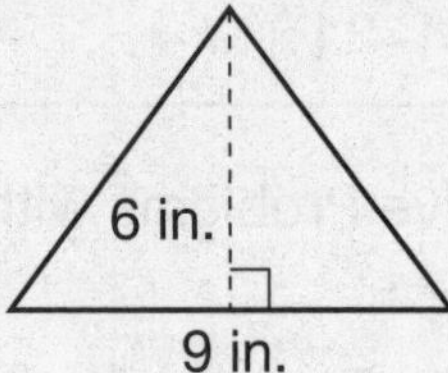

A. 54 square inches

B. 27 square inches

C. 24 square inches

D. 13.5 square inches

2. Which solid figure has 3 faces that are rectangles and 2 faces that are triangles?

A. triangular prism

B. triangular pyramid

C. square pyramid

D. rectangular prism

3. What is the area of this figure?

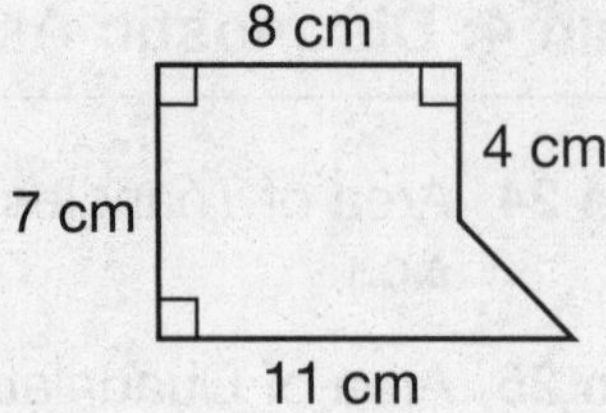

A. 60.5 cm^2

B. 62 cm^2

C. 65 cm^2

D. 68 cm^2

4. What is the volume of this box?

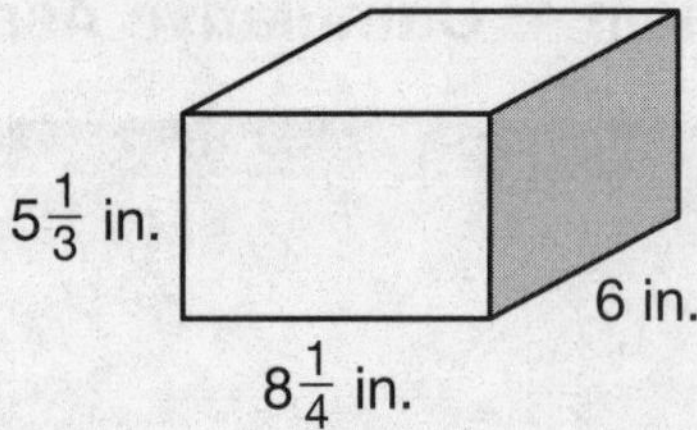

A. 240 $in.^3$

B. 264 $in.^3$

C. 324 $in.^3$

D. 528 $in.^3$

5. Which two solid figures have the same number of faces, edges, and vertices?

   A. rectangular prism and cube
   B. rectangular prism and rectangular pyramid
   C. triangular prism and triangular pyramid
   D. triangular pyramid and cube

6. Lakeisha measured the box shown below. What is the surface area of the box?

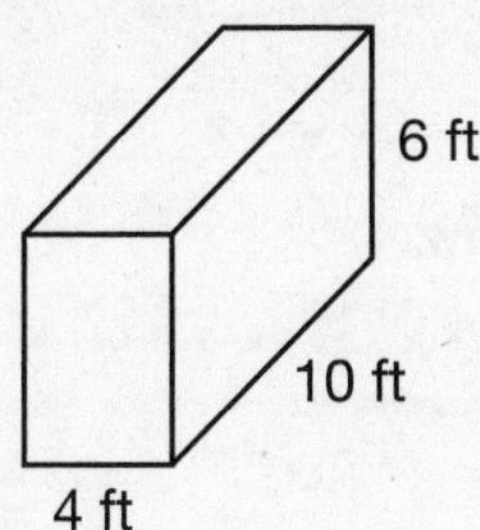

   A. 124 ft^2
   B. 224 ft^2
   C. 240 ft^2
   D. 248 ft^2

7. What is the perimeter of the rectangle on the coordinate plane below?

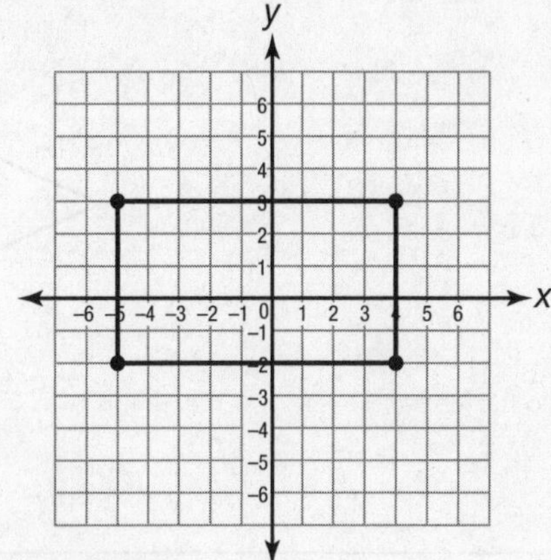

   A. 14 units
   B. 28 units
   C. 45 units
   D. 56 units

8. Phil has a storage shed in his yard measuring $6\frac{1}{2}$ feet by 9 feet by $10\frac{1}{2}$ feet. What is the volume of the storage shed?

   A. $274\frac{5}{8}$ cubic feet
   B. 540 cubic feet
   C. $614\frac{1}{4}$ cubic feet
   D. 693 cubic feet

9. The net of a square pyramid is shown below. What is the surface area of the pyramid?

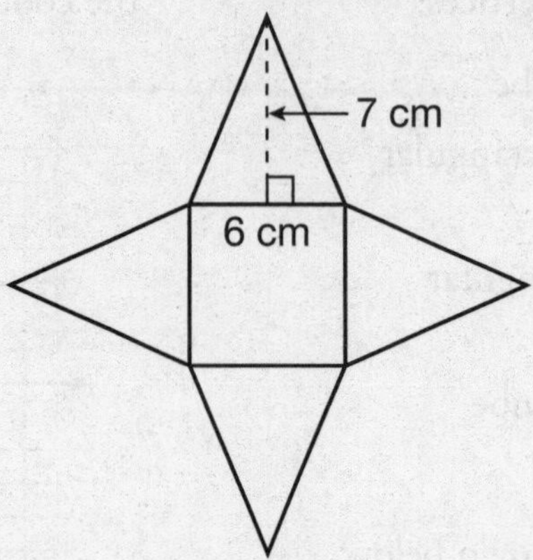

---

10. Julio and Britney each calculated the area of the trapezoid below.

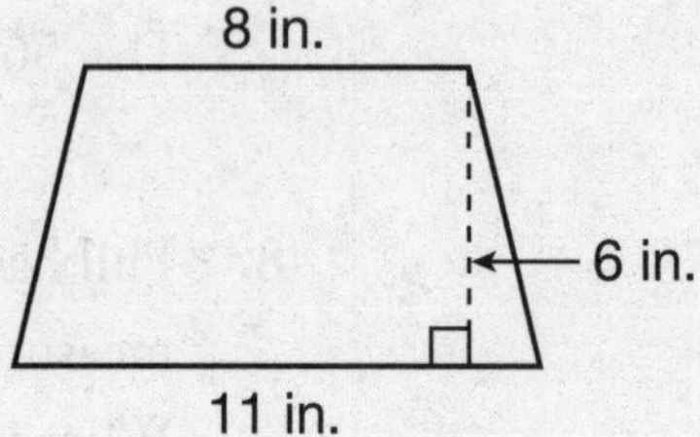

A. Julio used the formula for the area of a trapezoid to find the area. Show Julio's work.

B. Britney divided the trapezoid into two triangles to find the area. Show Britney's work.

C. What is the area of the trapezoid?

Common Core State Standard:
6.G.1

Area of Triangles

Getting the Idea

Area is a measure of the number of square units needed to cover a region. A **square unit** is a square with a side length of 1 of any particular unit. Square units can be square inches (in.^{2}), square centimeters (cm^2), or any other squared unit length.

The formula for the area of a triangle is $A = \frac{1}{2}bh$, where b represents the base length and h represents the height of the triangle. Some examples of triangles, with their bases and heights labeled, are shown below.

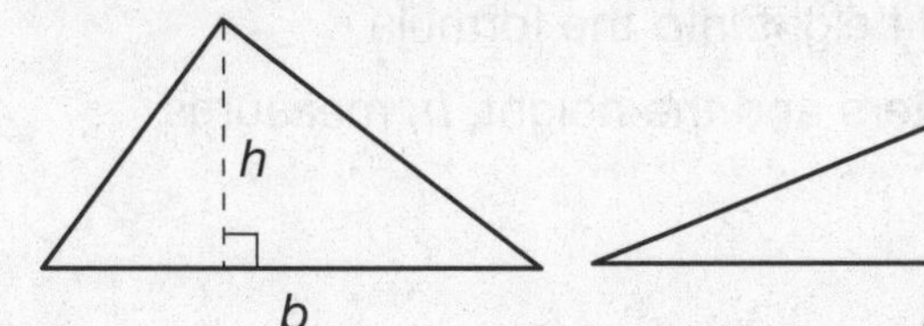

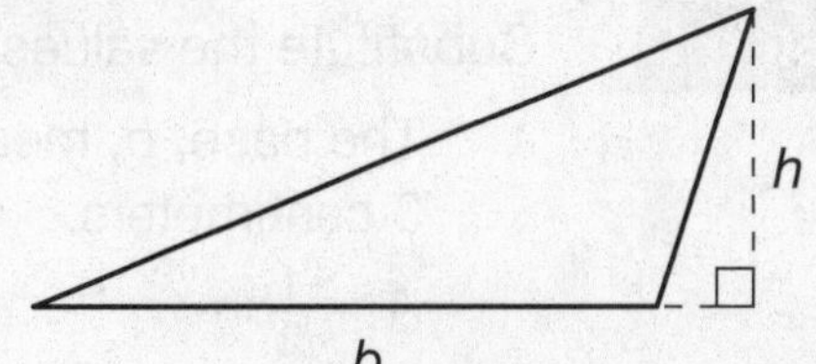

Example 1

What is the area of this triangle?

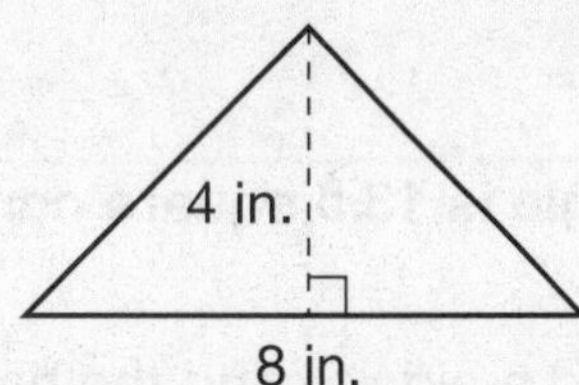

Strategy **Use the formula for the area of a triangle.**

Step 1 Substitute the values for the base and height into the formula.

The base, b, measures 8 inches and the height, h, measures 4 inches.

$A = \frac{1}{2}bh$

$A = \frac{1}{2} \times 8 \times 4$

Step 2 Multiply.

$A = \frac{1}{2} \times 8 \times 4$

$A = 4 \times 4$

$A = 16$

Solution **The area of the triangle is 16 square inches, or 16 in.^{2}**

Remember that a right triangle has a hypotenuse and 2 sides called legs. The legs form a right angle. To find the area of a right triangle, use the legs as the base and height.

Example 2

What is the area of this triangle?

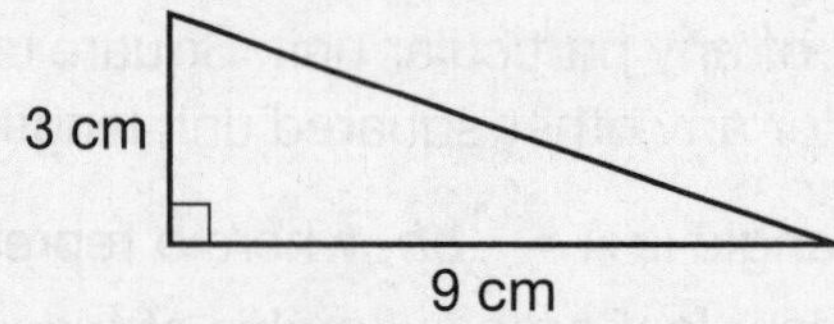

Strategy **Use the formula for the area of a triangle.**

Step 1 Substitute the values for the base and height into the formula.

The base, b, measures 9 centimeters and the height, h, measures 3 centimeters.

$A = \frac{1}{2}bh$

$A = 0.5 \times 9 \times 3$

Step 2 Multiply.

$A = 0.5 \times 9 \times 3$

$A = 4.5 \times 3$

$A = 13.5$

Solution **The area of the triangle is 13.5 square centimeters, or 13.5 cm^2.**

In an obtuse triangle, you can extend a side to find the height.

Example 3

What is the area of this triangle?

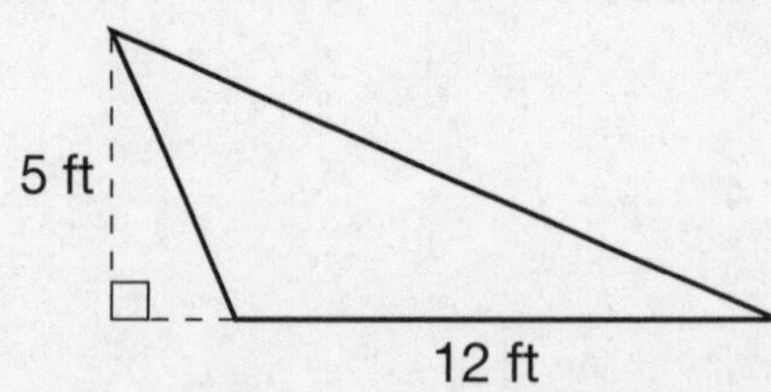

Strategy **Use the formula for the area of a triangle.**

Step 1 Substitute the values for the base and height into the formula.

The base, b, measures 12 feet and the height, h, measures 5 feet.

$A = \frac{1}{2}bh$

$A = \frac{1}{2} \times 12 \times 5$

Step 2 Multiply.

$A = \frac{1}{2} \times 12 \times 5$

$A = 6 \times 5$

$A = 30$

Solution **The area of the triangle is 30 square feet, or 30 ft^2.**

Coached Example

The Clemente family built a triangular deck at the back of their house, as shown below. What is the area of the Clementes' deck?

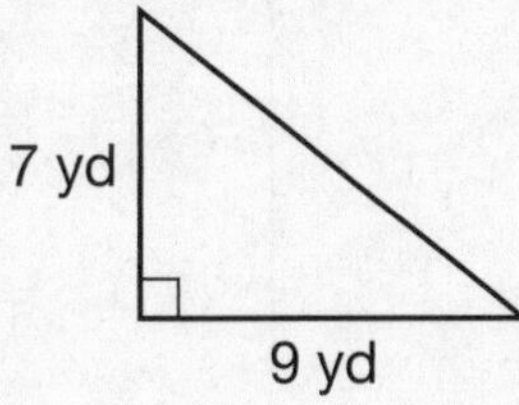

The deck is in the shape of a triangle.

The base of the triangle is _______ yards long and the height is _______ yards long.

The formula for the area of a triangle is $A =$ __________.

Substitute the values for the base and height into the formula.

$A = \frac{1}{2} \times$ _______ $\times$ _______

$A =$ _______ $\times$ _______

$A =$ _______

What units should be used to express the area? ________________

The area of the Clementes' deck is ____________________.

Lesson Practice

Choose the correct answer.

1. What is the area of this triangle?

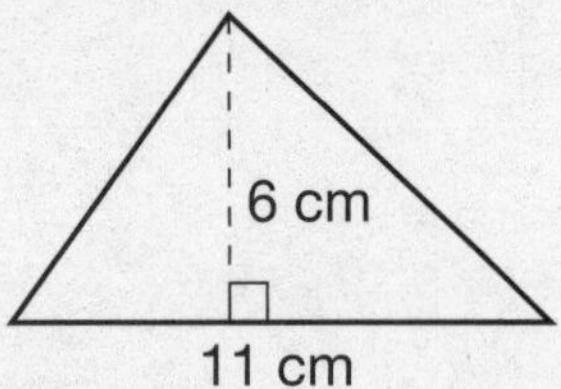

A. 16.5 cm^2

B. 17 cm^2

C. 33 cm^2

D. 66 cm^2

2. What is the area of this triangle?

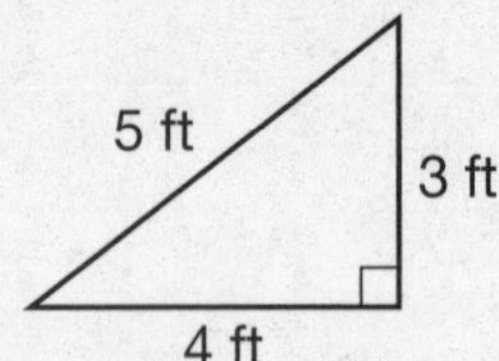

A. 6 ft^2

B. 12 ft^2

C. 13 ft^2

D. 22 ft^2

3. A triangular pennant has a base that is 9 inches long and a height of 19 inches. What is the area of the pennant?

A. 14 $in.^2$

B. 28 $in.^2$

C. $85\frac{1}{2}$ $in.^2$

D. 171 $in.^2$

4. What is the area of this triangle?

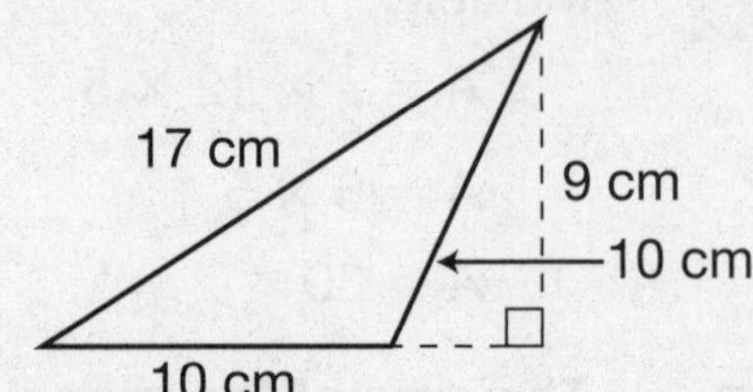

A. 45 cm^2

B. 46 cm^2

C. 50 cm^2

D. 85 cm^2

5. What is the area of this triangle?

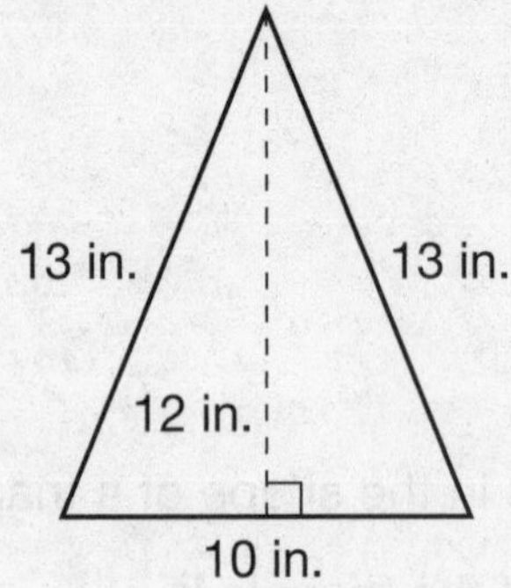

A. 39 $in.^2$

B. 60 $in.^2$

C. 65 $in.^2$

D. 120 $in.^2$

6. A flower bed in the shape of a right triangle has legs that measure 16 feet and 9 feet. What is the area of the flower bed?

A. 12.5 ft^2

B. 25 ft^2

C. 72 ft^2

D. 144 ft^2

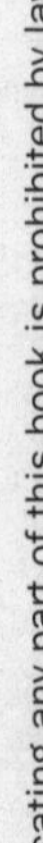

7. What is the area of this triangle?

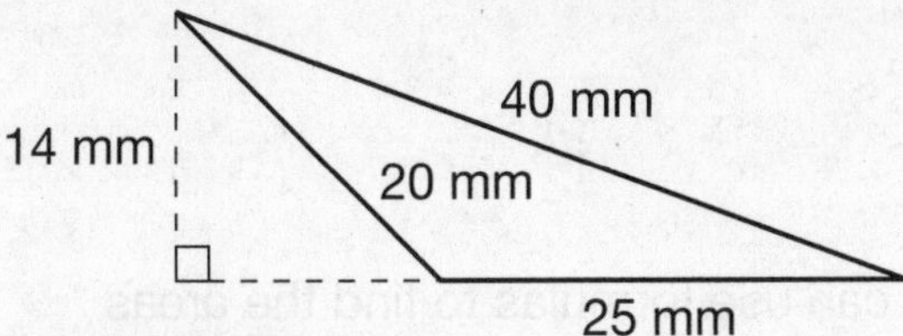

A. 140 mm^2

B. 175 mm^2

C. 200 mm^2

D. 400 mm^2

8. What is the area of this triangle?

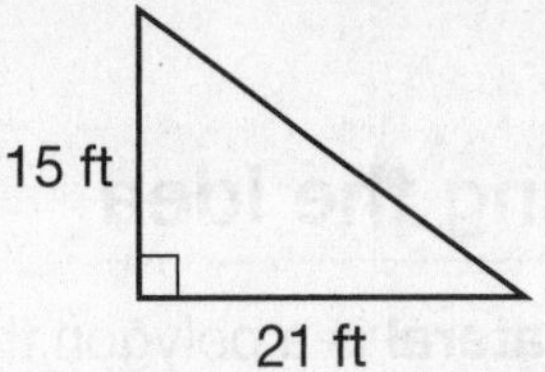

A. 315 ft^2

B. 305 ft^2

C. 285 ft^2

D. $157\frac{1}{2}ft^2$

9. Mrs. Green drew these two triangles on the board.

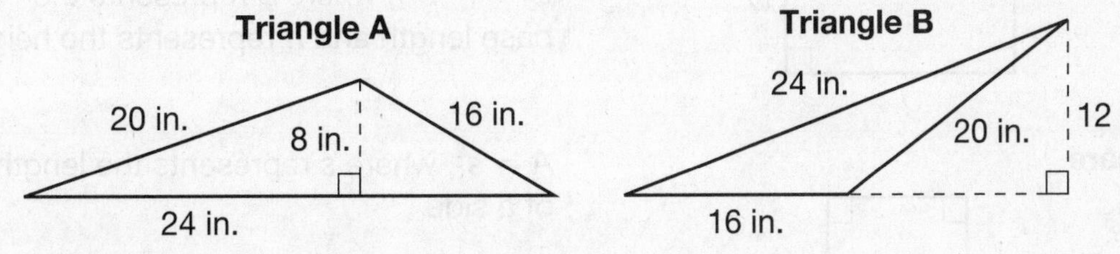

A. What is the area of triangle A? Show your work.

B. What is the area of triangle B? Show your work.

C. What do you notice about the two triangles? Explain your answer.

Common Core State Standard:
6.G.1

Area of Quadrilaterals

Getting the Idea

A **quadrilateral** is a polygon that has 4 sides. You can use formulas to find the areas of quadrilaterals.

Figure	Area Formula
Parallelogram (figure labeled h, b)	$A = bh$, where b represents the base length and h represents the height
Rectangle (figure labeled w, l)	$A = lw$, where l represents the length and w represents the height Or $A = bh$, where b represents the base length and h represents the height
Square (figure labeled s)	$A = s^2$, where s represents the length of a side
Rhombus (figure labeled h, b)	$A = bh$, where b represents the base length and h represents the height
Trapezoid (figure labeled b_1, h, b_2)	$A = \frac{1}{2}h(b_1 + b_2)$, where h represents the height and b_1 and b_2 represent the lengths of the bases

A diagonal of a rectangle divides the rectangle into two right triangles that are equal in area.

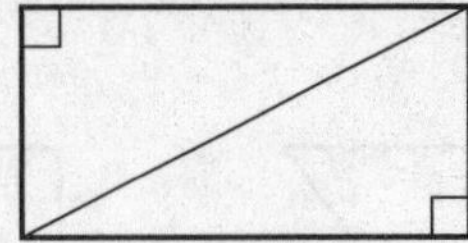

The diagram above shows that the area of a rectangle is equal to the area of the two triangles formed by the diagonal. Remember that the formula for the area of a triangle is $A = \frac{1}{2}bh$. So, the formula for the area of a rectangle is $A = 2 \times \frac{1}{2}bh$, or simply $A = bh$. You can substitute length for base and width for height to make the formula $A = lw$.

Example 1

What is the area of this rectangle?

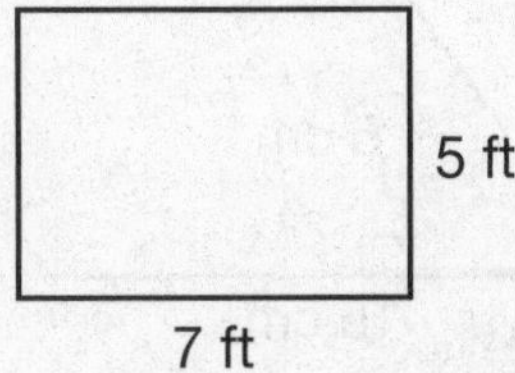

Strategy **Use the formula for the area of a rectangle.**

Step 1 Substitute the values for the length and width into the formula.

The length of the rectangle is 7 feet and the width is 5 feet.

$A = lw$

$A = 7 \times 5$

Step 2 Multiply.

$A = 7 \times 5 = 35$

Solution **The area of the rectangle is 35 square feet, or 35 ft^2.**

The formula for the area of a parallelogram, $A = bh$, is the same as the formula for the area of a rectangle. The diagram below will help you see why the same formula works for both kinds of quadrilaterals.

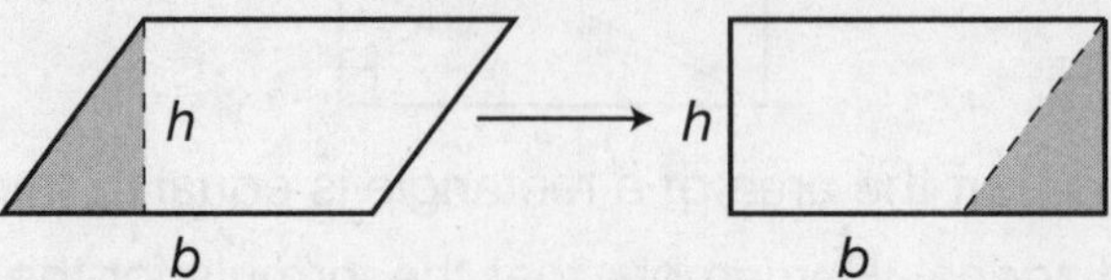

Moving the right triangle from the left side to the right side of the parallelogram forms a rectangle. The rectangle has the same area as the original parallelogram.

Example 2

What is the area of this parallelogram?

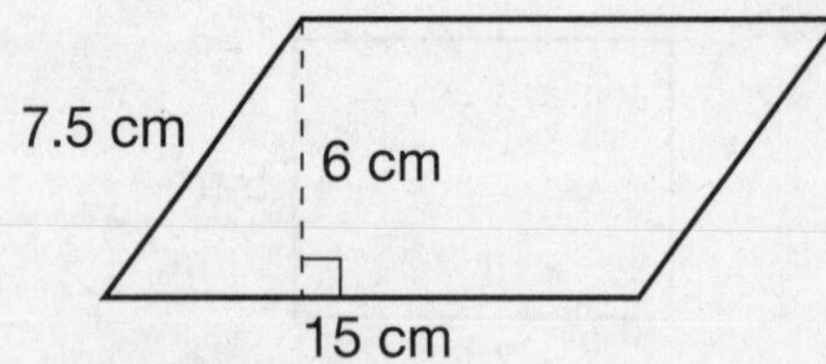

Strategy **Use the formula for the area of a parallelogram.**

Step 1 Substitute the values for the base and height into the formula.

The side length of 7.5 centimeters is not the height.

The base of the parallelogram is 15 cm and the height is 6 cm.

$A = bh$

$A = 15 \times 6$

Step 2 Multiply.

$A = 15 \times 6 = 90$

Solution **The area of the parallelogram is 90 square centimeters, or 90 cm^2.**

A rhombus is a parallelogram whose sides are all the same length. So, the formulas for the area of a rhombus and the area of a parallelogram are the same.

Example 3

What is the area of this rhombus?

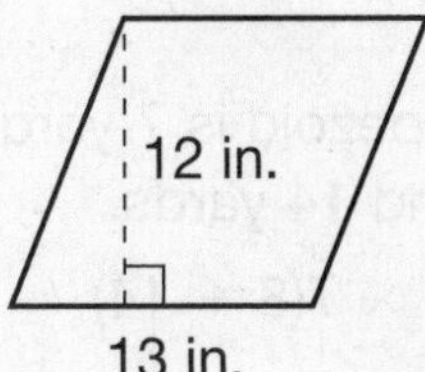

Strategy **Use the formula for the area of a rhombus.**

Step 1 Substitute the values for the base and height into the formula.

The base of the rhombus is 13 inches and the height is 12 inches.

$A = bh$

$A = 13 \times 12$

Step 2 Multiply.

$A = 13 \times 12 = 156$

Solution **The area of the rhombus is 156 square inches, or 156 in.^{2}.**

A diagonal of a trapezoid divides the trapezoid into two triangles, as shown below.

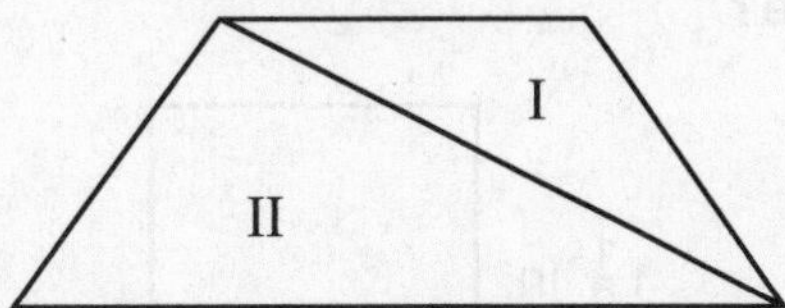

The area of the trapezoid is the sum of the areas of the triangles. Notice that the height is the same for both triangles.

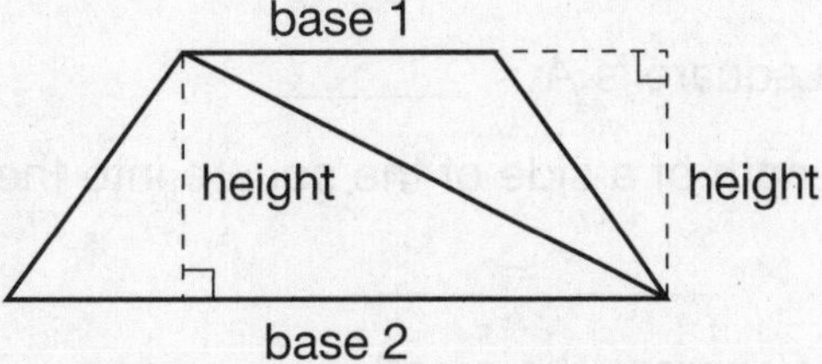

$A = \frac{1}{2}b_1h + \frac{1}{2}b_2h$

You can use the distributive property to rewrite the formula. The common factors in each term are $\frac{1}{2}$ and h.

$A = \frac{1}{2}b_1h + \frac{1}{2}b_2h = \frac{1}{2}h(b_1 + b_2)$

Example 4

What is the area of this trapezoid?

8 yd
7 yd
14 yd

Strategy **Use the formula for the area of a trapezoid.**

Step 1 Substitute the values for the bases and height into the formula.

The height of the trapezoid is 7 yards and the bases are 8 yards and 14 yards.

$A = \frac{1}{2}h(b_1 + b_2) = \frac{1}{2} \times 7(8 + 14)$

Step 2 Use the order of operations and number properties.

$A = \frac{1}{2} \times 7(8 + 14)$

$A = \frac{1}{2} \times 7 \times 22 = \frac{1}{2} \times 22 \times 7$

$A = 11 \times 7 = 77$

Solution **The area of the trapezoid is 77 square yards, or 77 yd².**

A **regular polygon** is a polygon in which all sides and all angles are congruent. A square is an example of a regular polygon.

Coached Example

What is the area of this square?

$1\frac{1}{2}$ in.

Each side of the square is ________ inches long.

The formula for the area of a square is A = ________.

Substitute the value of the length of a side of the square into the formula.

A = (________)² = ________ × ________ = ________

What units should be used to express the area? ________________

The area of the square is ________________________.

Lesson Practice

Choose the correct answer.

1. What is the area of this rhombus?

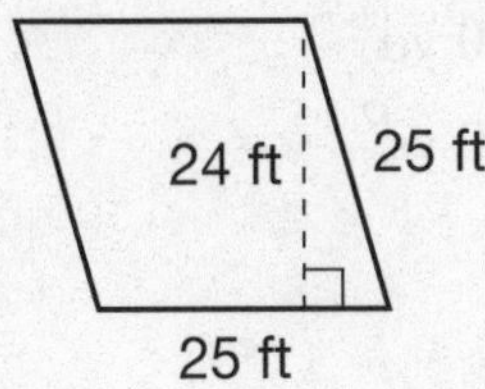

A. 100 ft^2
B. 576 ft^2
C. 600 ft^2
D. 625 ft^2

2. What is the area of this parallelogram?

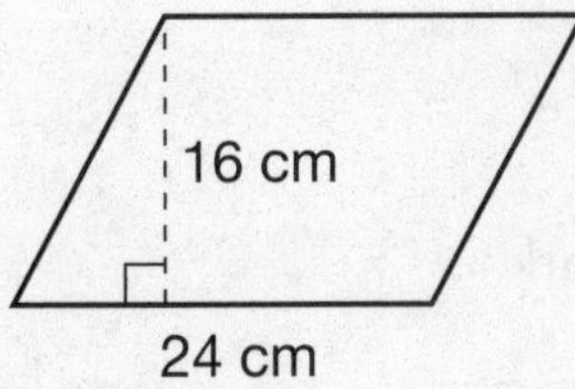

A. 288 cm^2
B. 384 cm^2
C. 504 cm^2
D. 576 cm^2

3. Nikki's bedroom is shaped like a rectangle that is 18 feet long and 12 feet wide. She wants to carpet the entire room. How many square feet of carpeting does she need?

A. 40 square feet
B. 80 square feet
C. 108 square feet
D. 216 square feet

4. What is the area of this trapezoid?

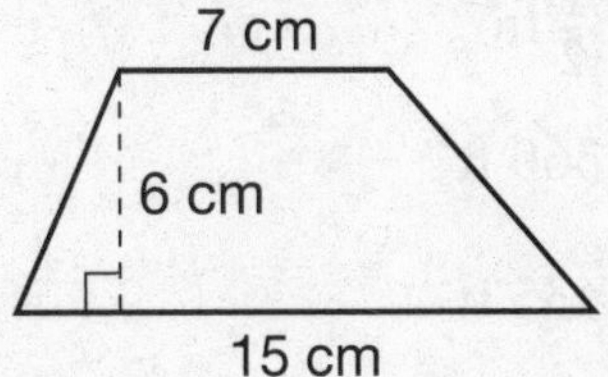

A. 28 cm^2
B. 66 cm^2
C. 132 cm^2
D. 630 cm^2

5. What is the area of this square?

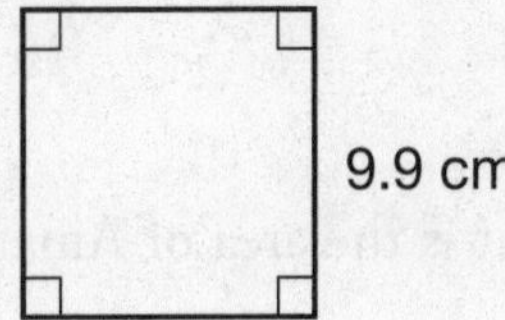

A. 3.3 cm^2
B. 39.6 cm^2
C. 81.81 cm^2
D. 98.01 cm^2

6. Lex built a rectangular pen outdoors for his dog Luther. The pen is 36 feet long and 27 feet wide. What is the area of the pen?

A. 126 ft^2
B. 486 ft^2
C. 972 ft^2
D. 3,969 ft^2

7. Mae's rose garden is in the shape of a trapezoid with a height of 35 feet. The bases of the garden measure 50 feet and 32 feet. What is the area of Mae's rose garden?

A. $58\frac{1}{2}$ ft^2

B. 1,360 ft^2

C. 1,435 ft^2

D. 2,870 ft^2

8. The schoolyard at Kenny's school is a square that is 50 yards long on each side. What is the area of the schoolyard?

A. 2,500 yd^2

B. 2,000 yd^2

C. 250 yd^2

D. 200 yd^2

9. Amanda drew the trapezoid shown below.

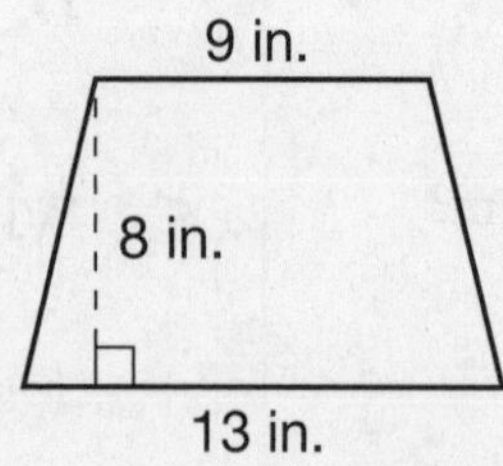

A. What is the area of Amanda's trapezoid? Show your work.

B. Explain how the formula for the area of a triangle can be used to find the area of Amanda's trapezoid.

Common Core State Standard:
6.G.1

Area of Composite Polygons

Getting the Idea

A **composite polygon** is a polygon that can be divided into simpler figures. To find the area of a composite polygon, find the sum of the areas of the simpler figures.

The formulas in the table below will help you find the area of many composite polygons.

Figure	Area Formula
Rectangle	$A = lw$
Square	$A = s^2$
Triangle	$A = \frac{1}{2}bh$

Example 1

Eric drew a diagram of his back yard. What is the area of his back yard?

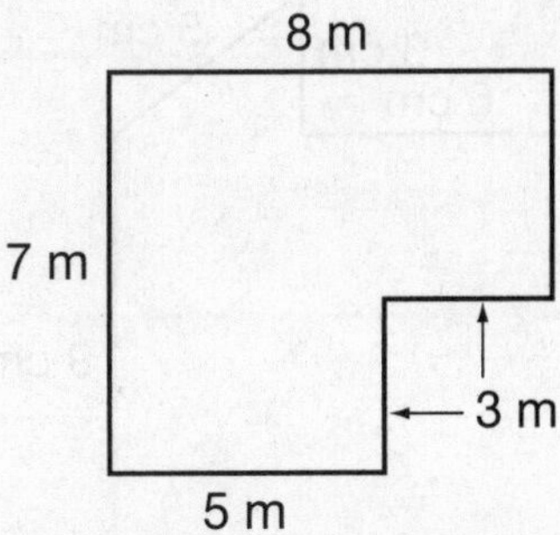

Strategy **Divide the figure into smaller figures. Add the areas of the smaller figures.**

Step 1 Divide the figure into two rectangles.

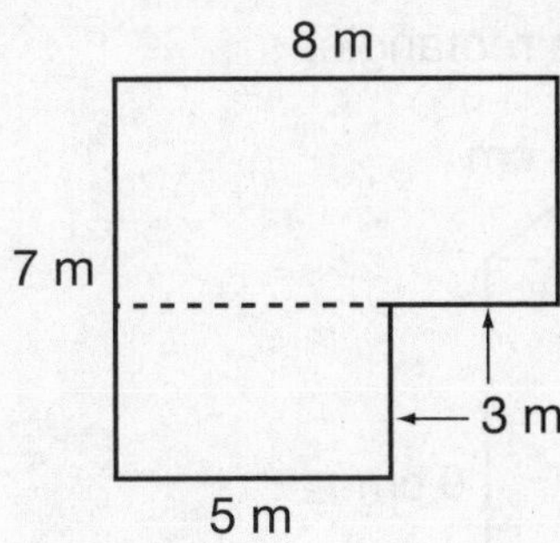

Step 2 Subtract to find the width of the larger rectangle.

$7 - 3 = 4$

The larger rectangle has a length of 8 meters and a width of 4 meters.

Step 3 Find the area of the larger rectangle.

$A = lw = 8 \times 4 = 32$

The area of the larger rectangle is 32 square meters.

Step 4 Find the area of the smaller rectangle.

$A = lw = 5 \times 3 = 15$

The area of the smaller rectangle is 15 square meters.

Step 5 Add the areas.

$32 + 15 = 47$

Solution **The area of Eric's back yard is 47 square meters, or 47 m^2.**

Example 2

What is the area of this figure?

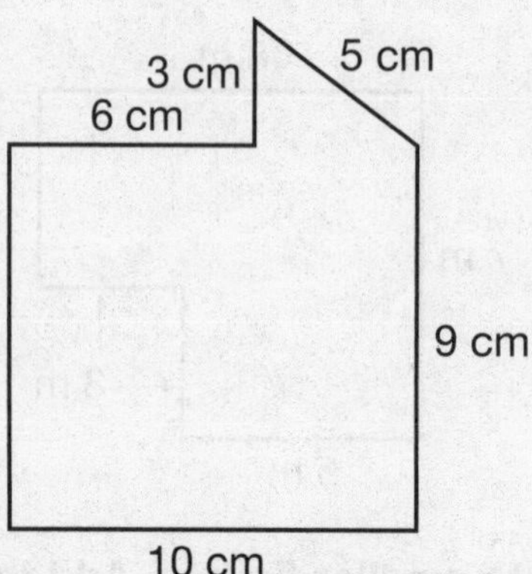

Strategy **Divide the figure into smaller figures. Add the areas of the smaller figures.**

Step 1 Divide the figure into a triangle and a rectangle.

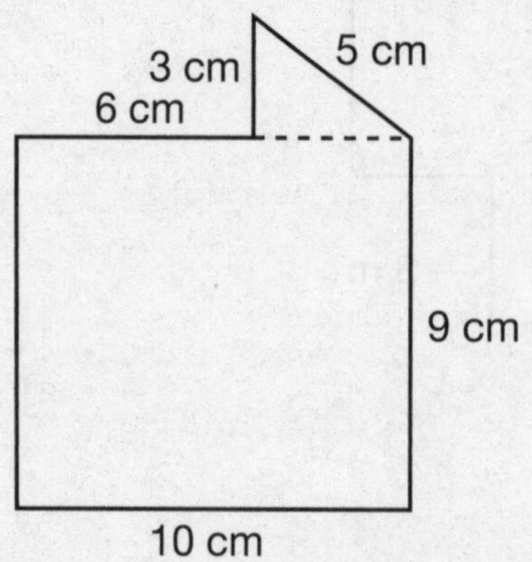

Step 2 Find the area of the rectangle.

$A = lw = 10 \times 9 = 90$

Step 3 Find the area of the triangle.

The length of the base is 4 centimeters since $10 - 6 = 4$.

$A = \frac{1}{2}bh = \frac{1}{2} \times 4 \times 3 = 6$

Step 4 Add the areas.

$90 + 6 = 96$

Solution **The area of the figure is 96 square centimeters, or 96 cm^2.**

Coached Example

What is the area of this figure?

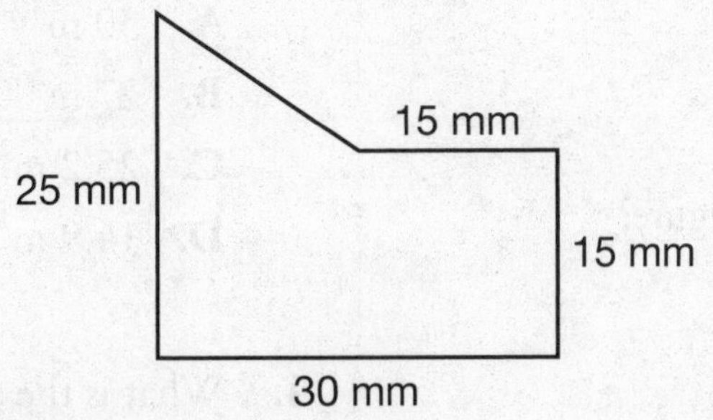

Divide the figure into a rectangle and a triangle.

The dimensions of the rectangle are __________ by __________.

The formula for the area of a rectangle is A = __________.

A = __________ × __________ = __________

The area of the rectangle is __________ square millimeters.

In millimeters, the base of the triangle is 30 − __________ = __________.

In millimeters, the height of the triangle is 25 − __________ = __________.

The formula for the area of a triangle is A = __________.

A = __________ × __________ × __________ = __________

The area of the triangle is __________ square millimeters.

Add the areas. __________ + __________ = __________

The area of the figure is __________ square millimeters.

Lesson Practice

Choose the correct answer.

1. What is the area of this figure?

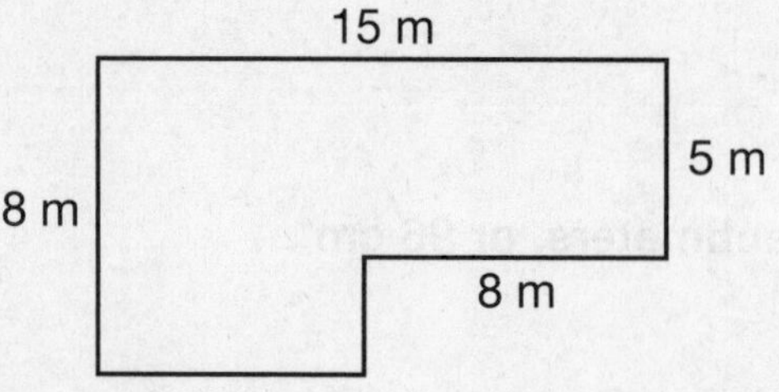

A. 106 m^2

B. 96 m^2

C. 56 m^2

D. 36 m^2

2. What is the area of this figure?

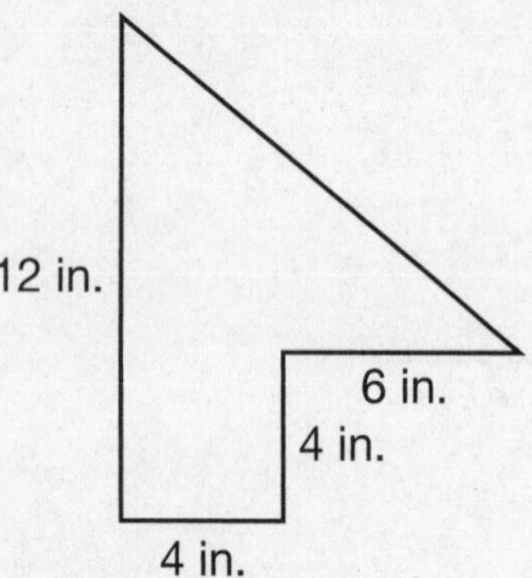

A. 52 $in.^2$

B. 56 $in.^2$

C. 76 $in.^2$

D. 96 $in.^2$

3. Mr. Blackburn is buying new carpet for his family room and hallway. The floor plan is shown below.

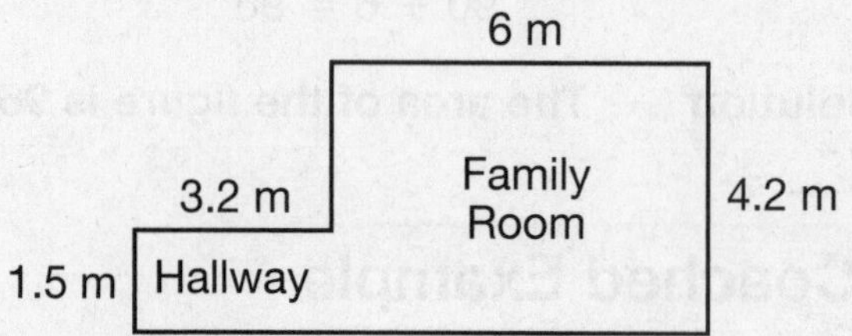

How much carpet does he need?

A. 30 m^2

B. 27 m^2

C. 25.2 m^2

D. 14.9 m^2

4. What is the area of this figure?

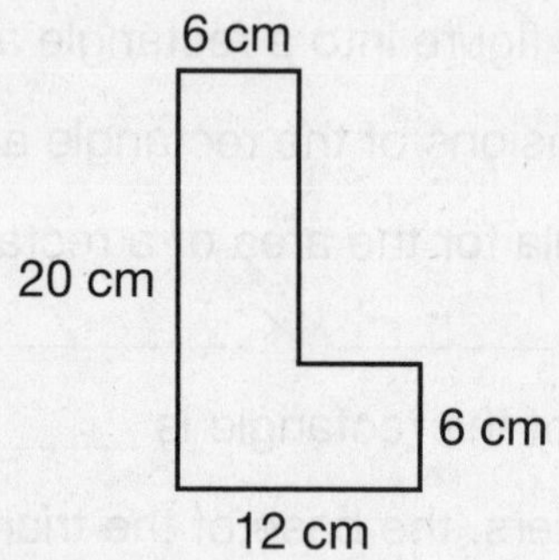

A. 240 cm^2

B. 192 cm^2

C. 156 cm^2

D. 40 cm^2

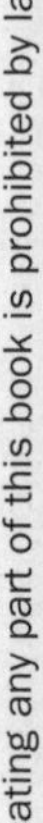

5. What is the area of this figure?

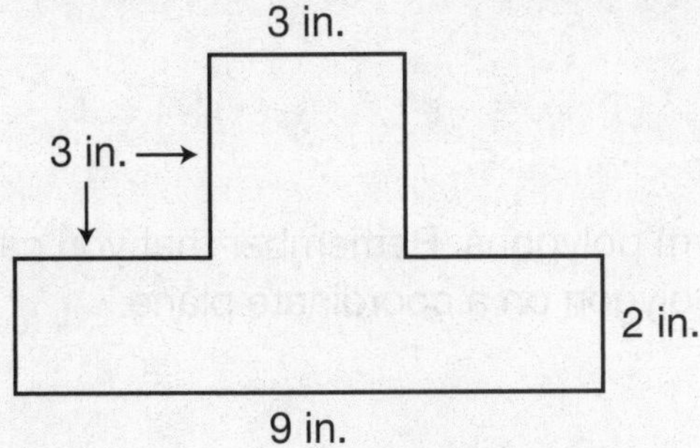

A. 25 in.^{2}
B. 27 in.^{2}
C. 28 in.^{2}
D. 30 in.^{2}

6. What is the area of this figure?

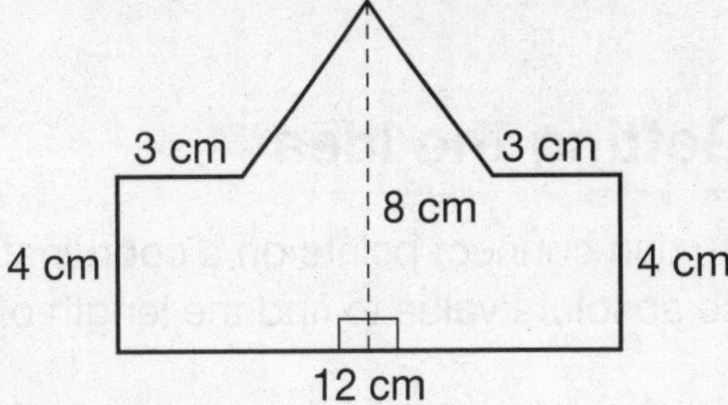

A. 96 cm^2
B. 80 cm^2
C. 72 cm^2
D. 60 cm^2

7. Use this figure to answer the questions below.

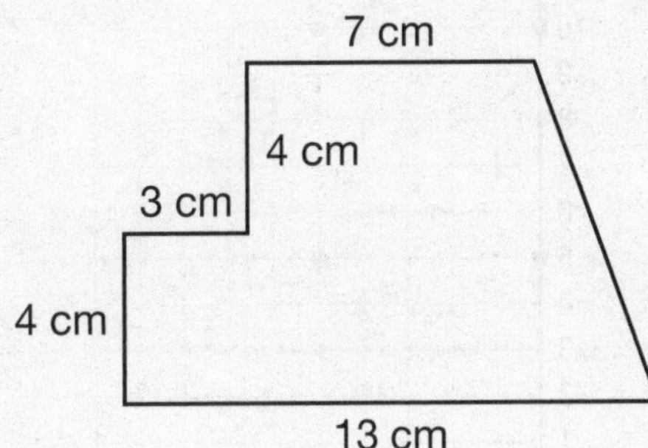

A. What is the area of the figure?

B. Explain how you found your answer to part A.

Common Core State Standard:
6.G.3

Polygons on the Coordinate Plane

Getting the Idea

You can connect points on a coordinate plane to form polygons. Remember that you can use absolute value to find the length of a side of a polygon on a coordinate plane.

Example 1

Plot the points (5, 10), (5, 5), (0, 5), and (0, 10) on a coordinate grid. Then connect the points. What figure is formed?

Strategy **Plot the points. Then connect the points to identify the figure formed.**

Step 1 Plot each point on the grid.

Plot each point.

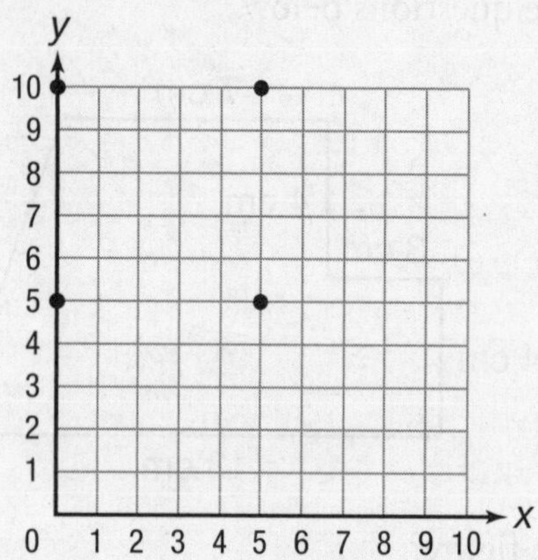

Step 2 Connect the points.

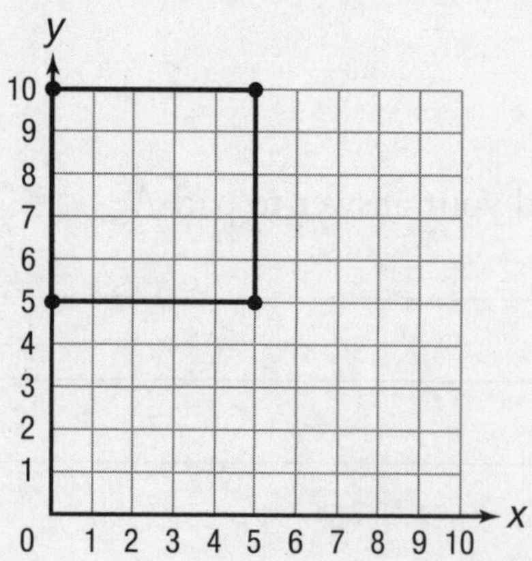

Step 3 Identify the geometric figure.

Count the number of units between the points on each vertical and horizontal side. Each side is 5 units long, and $|5| = 5$.

Each side of the figure is the same length. The figure is a square.

Solution **The points form a square.**

Example 2

John drew the floor plan for a new shed on a coordinate grid. He connected the points (2, 1), (2, 8), (5, 8), and (5, 1). Identify the figure formed by the connected points. Then find the perimeter, in yards, of the shed.

Strategy **Plot and connect the points to identify the shape of the floor plan. Then find the perimeter.**

Step 1 Plot each point on the grid. Then connect the points.

Draw lines to connect the points in order.

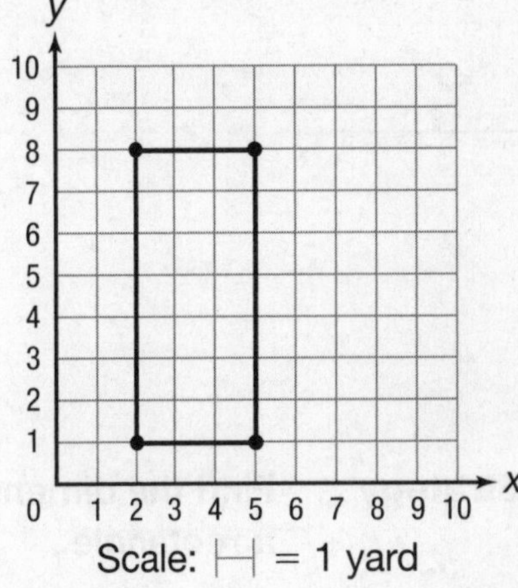

Step 2 Identify the figure.

The figure is a rectangle.

Step 3 Find the length of each side of the rectangle.

Count the number of units between two points on either vertical side.

The distance from (5, 1) to (5, 8) is 7 units, and $|7| = 7$.

Count the number of units between two points on either horizontal side.

The distance from (2, 8) to (5, 8) is 3 units, and $|3| = 3$.

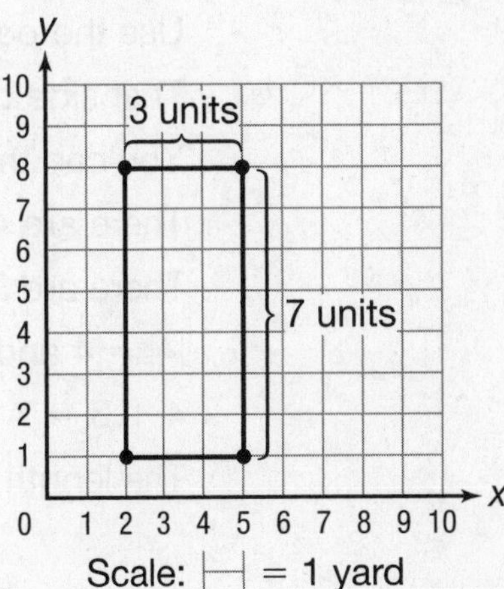

Step 4 Find the perimeter of the rectangle.

$P = 7 + 3 + 7 + 3 = 20$ units

Step 5 Interpret the scale.

The scale says that each unit represents 1 yard.

Since the perimeter of the figure on the grid is 20 units, the perimeter of the shed is 20 yards.

Solution **The perimeter of the shed is 20 yards.**

Example 3

Delilah is designing a rectangular kennel. She drew the plans on the coordinate grid below. What will be the area of the kennel?

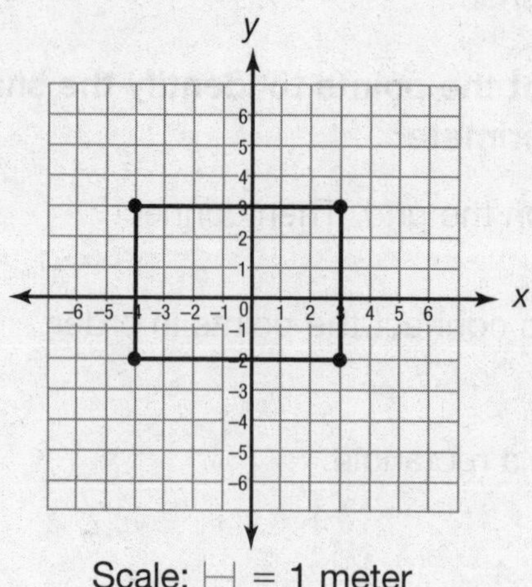

Scale: ⊢⊣ = 1 meter

Strategy **Find the dimensions of the rectangle. Use the formula for the area of a rectangle.**

Step 1 Find the length of the rectangle.

Use the points $(-4, -2)$ and $(3, -2)$ to find the length of the rectangle.

That side of the rectangle is part of the line $y = -2$.

You can think of $y = -2$ as a number line on which the y-axis is 0.

There are 4 units from $(-4, -2)$ to the y-axis on the line $y = -2$.

There are 3 units from $(3, -2)$ to the y-axis on the line $y = -2$.

$|4| = 4$ and $|3| = 3$

$4 + 3 = 7$

The length of the rectangle measures 7 units.

Step 2 Find the width of the rectangle.

Use the points $(-4, -2)$ and $(-4, 3)$ to find the width of the rectangle.

That side of the rectangle is part of the line $x = -4$.

You can think of $x = -4$ as a number line on which the x-axis is 0.

There are 2 units from $(-4, -2)$ to the x-axis on the line $x = -4$.

There are 3 units from $(-4, 3)$ to the x-axis on the line $x = -4$.

$|2| = 2$ and $|3| = 3$

$2 + 3 = 5$

The width of the rectangle measures 5 units.

Step 3 Use the formula for the area of a rectangle.

$A = lw$

$A = 7 \times 5$

$A = 35$

The area of the rectangle on the grid is 35 square units.

Step 4 Interpret the scale.

The scale says that each unit represents 1 meter.

Since the area of the figure on the grid is 35 square units, the area of the kennel will be 35 square meters.

Solution **The area of the kennel is 35 square meters.**

Coached Example

The figure below is the design for a new public swimming pool. What will be the area of the bottom of the pool?

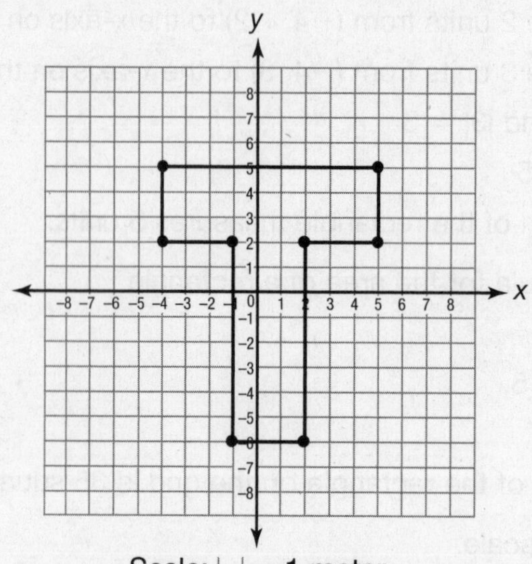

Divide the figure into two rectangles by drawing a horizontal line segment from point ________ to point ____________.

Label the top rectangle I and the bottom rectangle II.

Use absolute value to find the dimensions of each rectangle.

The length of rectangle I is ______ + ______ = ______ units.

The width of rectangle I is ______ units.

Use the area formula to find the area of rectangle I.

Area of rectangle I = ______ × ______ = ______ square units

The length of rectangle II is ______ + ______ = ______ units.

The width of rectangle II is ______ + ______ = ______ units.

Use the area formula to find the area of rectangle II.

Area of rectangle II = ______ × ______ = ______ square units

Add the areas of the rectangles.

Area of rectangle I + area of rectangle II = ______ + ______ = ______ square units

Based on the scale, each unit equals 1 ____________.

The area of the bottom of the pool will be ______ square meters.

Lesson Practice

Choose the correct answer.

Use the coordinate grid for questions 1 and 2.

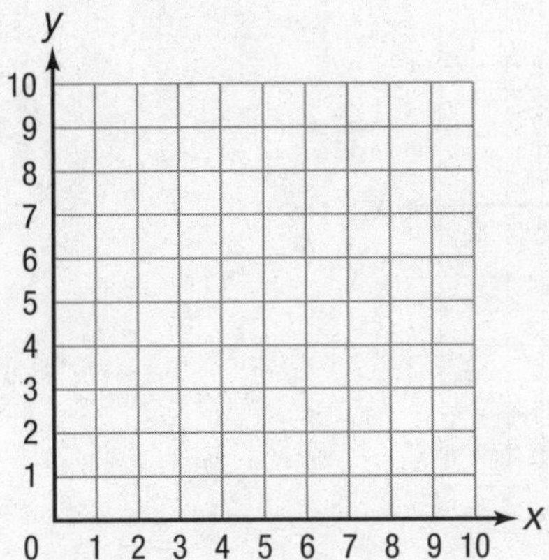

1. Draw and connect the points (1, 9), (4, 9), (4, 7), and (1, 7). Which best describes the kind of geometric figure that is formed?

   A. square
   B. rhombus
   C. rectangle
   D. pentagon

2. Draw and connect the points (5, 1), (6, 4), (9, 4), and (10, 1). Which best describes the kind of geometric figure that is formed?

   A. rhombus
   B. trapezoid
   C. rectangle
   D. hexagon

Use the coordinate grid and the information below for questions 3 and 4.

Debra drew designs for 3 different tables on the grid below.

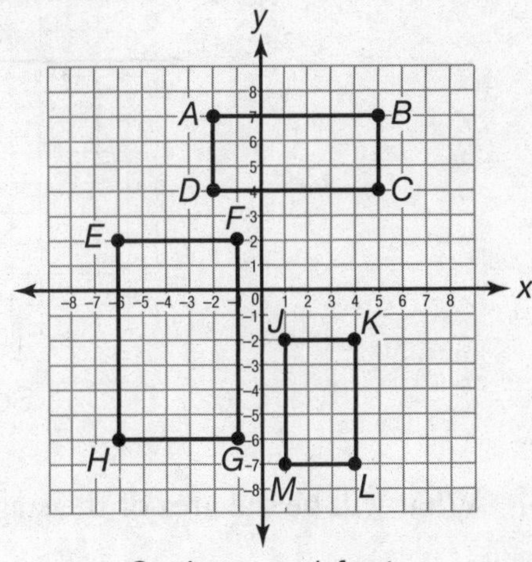

Scale: ⊢ = 1 foot

3. What will be the perimeter of the table represented by figure *EFGH*?

   A. 13 feet
   B. 22 feet
   C. 26 feet
   D. 40 feet

4. How much greater will the area of the table represented by figure *ABCD* be than that of the table represented by figure *JKLM*?

   A. 6 square feet
   B. 19 square feet
   C. 27 square feet
   D. 40 square feet

5. Allan drew a floor plan for a theater's stage on a coordinate plane. He used the points (−3, 4), (6, 4), (6, −1), and (−3, −1).

A. Plot and connect the points he used on the coordinate plane below.

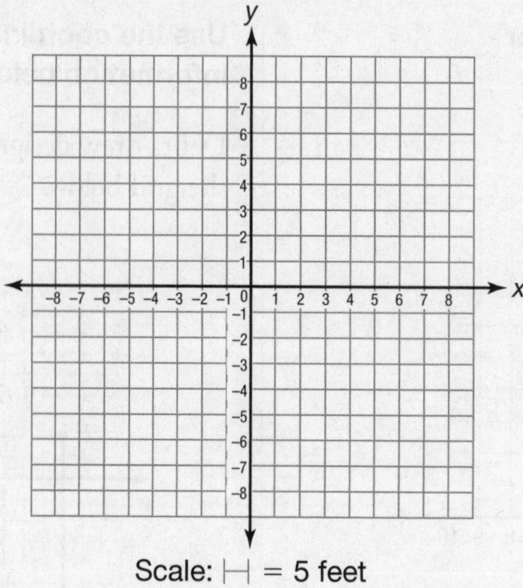

Scale: ⊢⊣ = 5 feet

B. What will be the area of the stage?

____

C. Explain how you found the area of the stage.

____

____

____

Common Core State Standard:
6.G.1

Solve Problems with Area

Getting the Idea

You can use the area formulas to solve problems.

Example 1

Nikki made a flag shaped like a triangle out of red cloth. The triangle has a base of 18 inches and a height of 12 inches. How many square inches of cloth did she use?

Strategy **Use the formula for the area of a triangle.**

Step 1 Substitute the values for the base and height into the formula.

The base of the triangle is 18 inches long and the height is 12 inches.

$A = \frac{1}{2}bh$

$A = \frac{1}{2} \times 18 \times 12$

Step 2 Multiply.

$A = \frac{1}{2} \times 18 \times 12 = 9 \times 12 = 108$

Solution **Nikki used 108 square inches of cloth.**

Example 2

Alan's kitchen is in the shape of a square that is 8 feet long on each side. He wants to cover the floor with 1-foot by 1-foot tiles. How many tiles does he need to cover the floor?

Strategy **Use the formula for the area of a square.**

Step 1 Substitute the value for the length of a side into the formula.

The length of a side of the square is 8 feet.

$A = s^2$

$A = 8^2$

Step 2 Evaluate.

$A = 8^2 = 8 \times 8 = 64$

Step 3 Determine how many tiles are needed.

The area of the floor is 64 square feet. The area of each tile is 1×1, or 1 square foot. So, Alan will need 64 tiles.

Solution **Alan needs 64 1-foot by 1-foot tiles to cover the floor.**

Example 3

The Sanfords are getting new carpet for a bedroom and a hallway in their house. The bedroom is 12 feet long and 10 feet wide. The hallway is 20 feet long and 4 feet wide. How much more carpeting do the Sanfords need for the bedroom than for the hallway?

Strategy **Find the areas of the bedroom and the hallway.**

Step 1 Identify the values of length, l, and width, w, for the bedroom.

$l = 12$ and $w = 10$

Step 2 Substitute the values into the formula for the area of a rectangle. Solve.

$A = lw$

$A = 12 \times 10 = 120$ square feet

Step 3 Identify the values of length, l, and width, w, for the hallway.

$l = 20$ and $w = 4$

Step 4 Substitute the values into the formula for the area of a rectangle. Solve.

$A = lw$

$A = 20 \times 4 = 80$ square feet

Step 5 Find the difference of the areas.

$120 - 80 = 40$

Solution **The Sanfords need 40 square feet more carpeting for the bedroom than for the hallway.**

Coached Example

Carrie needs to paint her garage door. The dimensions of the door are shown below.

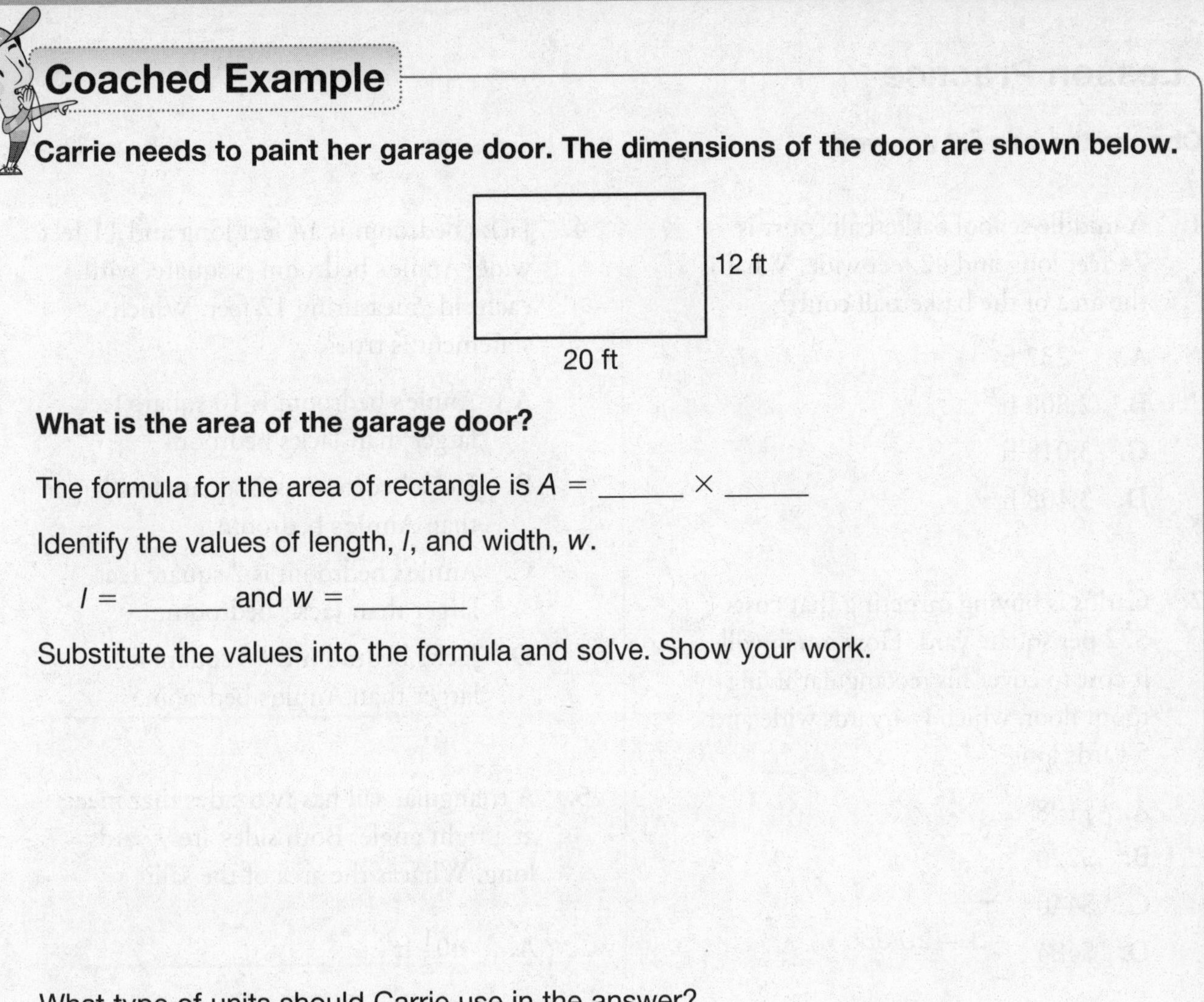

What is the area of the garage door?

The formula for the area of rectangle is $A =$ ______ × ______

Identify the values of length, *l*, and width, *w*.

$l =$ ______ and $w =$ ______

Substitute the values into the formula and solve. Show your work.

What type of units should Carrie use in the answer? ______________

The area of the garage door is ________________.

Lesson Practice

Choose the correct answer.

1. A middle school basketball court is 74 feet long and 42 feet wide. What is the area of the basketball court?

A. 232 ft^2

B. 2,808 ft^2

C. 3,018 ft^2

D. 3,108 ft^2

2. Carlos is buying carpeting that costs $22 per square yard. How much will it cost to cover his rectangular living room floor, which is 4 yards wide and 5 yards long?

A. $198

B. $220

C. $440

D. $484

3. Jessica wants to seed her backyard with grass seed. Her yard is 120 feet long and 90 feet wide. If Jessica buys bags of seed that cover 600 square feet, how many bags of seed will she need?

A. 5

B. 18

C. 21

D. 45

4. Jack's bedroom is 14 feet long and 11 feet wide. Annie's bedroom is square, with each side measuring 12 feet. Which statement is true?

A. Annie's bedroom is 10 square feet larger than Jack's bedroom.

B. Jack's bedroom is 2 square feet larger than Annie's bedroom.

C. Annie's bedroom is 2 square feet larger than Jack's bedroom.

D. Jack's bedroom is 10 square feet larger than Annie's bedroom.

5. A triangular sail has two sides that meet at a right angle. Both sides are 9 yards long. What is the area of the sail?

A. $40\frac{1}{2}$ ft^2

B. 45 ft^2

C. 81 ft^2

D. 162 ft^2

6. A commemorative plaque is in the shape of a trapezoid with a height of 8 inches and bases that measure 12 inches and 15 inches. What is the area of the plaque?

A. 108 $in.^2$

B. 135 $in.^2$

C. 150 $in.^2$

D. 163 ft^2

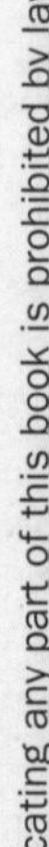

7. Patel is covering a rectangular-shaped trivet with 1-inch square tiles that cost \$0.15 each. The base of the trivet is 14 inches long and its height is 9 inches. How much will it cost to tile the trivet?

A. \$1.89

B. \$1.95

C. \$18.90

D. \$19.50

8. A suncatcher has 6 sections. Each section is in the shape of a parallelogram with a base of 12 cm and a height of 8 cm. What is the total area of the sections?

A. 768 cm^2

B. 576 cm^2

C. 432 cm^2

D. 120 cm^2

---

9. Brett has a square vegetable garden that measures 18 ft on each side. One bag of fertilizer can cover 54 square feet.

A. What is the area of the vegetable garden? Show your work.

B. How many bags of fertilizer will Brett need to cover the entire garden? Explain how you found your answer.

____

____

____

Common Core State Standard:
6.G.4

Solid Figures

Getting the Idea

Solid figures, also called **three-dimensional figures**, are figures that have length, width, and height.

Solid figures can be classified by the number of faces, edges, and vertices they have. A **face** is a flat surface of a solid figure. An **edge** is a line segment where two faces of a solid figure meet. A **vertex** is the point where three or more edges of a solid figure meet. The plural of vertex is vertices.

A **net** is flat pattern that can be folded into a three-dimensional figure. A net shows each surface of the solid figure it forms.

A **prism** is a three-dimensional figure with a pair of parallel faces called **bases** that are congruent polygons. Its other faces are rectangles or parallelograms. The table below shows some common prisms and their nets.

Cube	Rectangular Prism	Triangular Prism
face vertex edge		
6 faces 12 edges 8 vertices	6 faces 12 edges 8 vertices	5 faces 9 edges 6 vertices

A **pyramid** has a base that is a polygon. All its other faces are triangles. The table below shows some common pyramids and their nets.

Rectangular Pyramid	Triangular Pyramid
base	
5 faces 8 edges 5 vertices	4 faces 6 edges 4 vertices

Example 1

What kind of solid figure is shown?

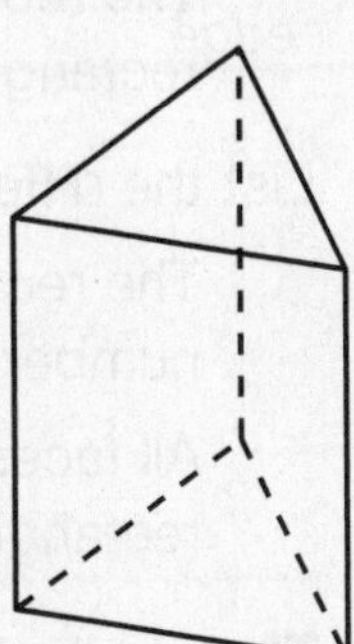

Strategy **Use the faces to identify the figure.**

Step 1 Count the number of faces.

There are 5 faces.

Triangular prisms and rectangular pyramids have 5 faces.

Step 2 Identify the shapes of the faces.

There are 3 rectangular faces and 2 triangular faces.

A rectangular pyramid does not have more than 1 rectangular face.

Solution **The figure is a triangular prism.**

Example 2

How are a rectangular prism and a rectangular pyramid alike? How are they different?

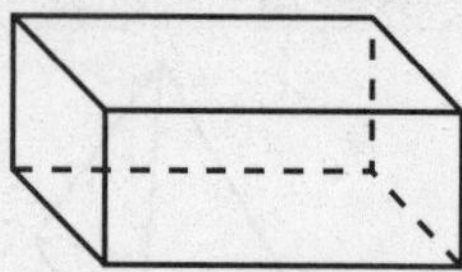
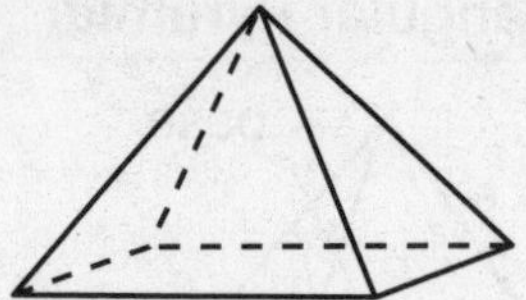

Strategy **Describe the properties of a rectangular prism and a rectangular pyramid.**

Step 1 Describe the number of faces in each figure.

A rectangular prism has 6 faces.

A rectangular pyramid has 5 faces.

Step 2 Describe the bases of each figure.

A rectangular prism has two parallel bases that are congruent rectangles.

A rectangular pyramid has one base that is a rectangle.

Step 3 Describe the faces of each figure.

The faces of a rectangular prism are rectangles.

The faces of a rectangular pyramid are triangles.

Step 4 List the similarities of the figures.

The rectangular prism and the rectangular pyramid each have a rectangular base.

Step 5 List the differences in the figures.

The rectangular prism and the rectangular pyramid have different numbers of faces.

All faces of the rectangular prism are rectangles. Only one face of the rectangle pyramid is a rectangle. All its other faces are triangles.

Solution **The similarities and differences in a rectangular prism and a rectangular pyramid are listed in Steps 4 and 5.**

Coached Example

How are a triangular prism and a triangular pyramid alike? How are they different?

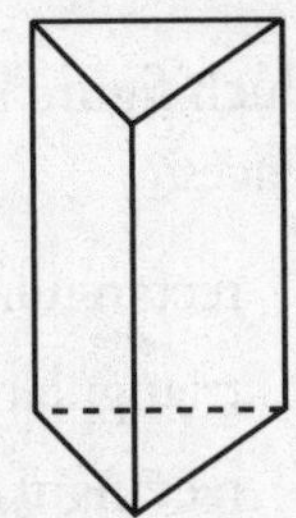

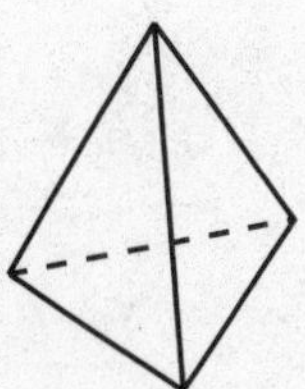

A triangular prism has _____ faces.

A triangular pyramid has _____ faces.

A triangular prism has two bases that are ________________.

A triangular pyramid has one base that is a ________________.

The faces of a triangular prism are ________________ and ________________.

The faces of a triangular pyramid are all ________________.

List the similarities in the figures.

A triangular prism and a triangular pyramid each have a ________________ base.

List the differences in the figures.

A triangular prism has _____ faces, but a triangular pyramid has _____ faces.

The faces of a triangular prism are ________________ and ________________, but the faces of a triangular pyramid are all ________________.

Lesson Practice

Choose the correct answer.

1. Which solid figure has 6 faces?

A.

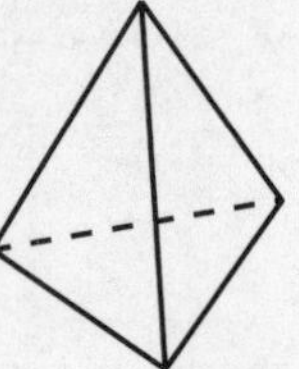

B.

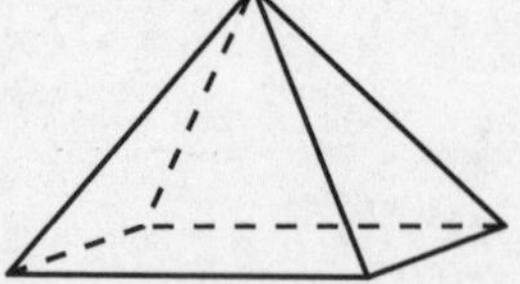

C.

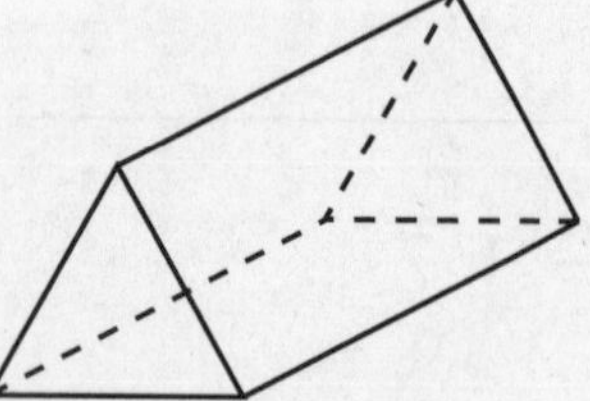

D.

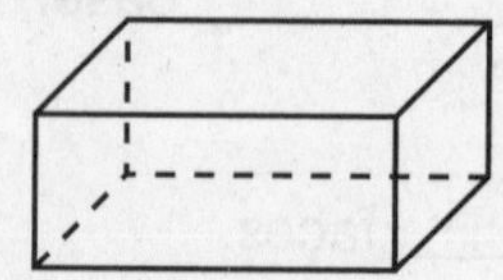

2. Which solid figure has only one base?

A. rectangular pyramid

B. cube

C. rectangular prism

D. triangular prism

3. Which figure has twice as many edges as faces?

A. rectangular pyramid

B. triangular pyramid

C. rectangular prism

D. triangular prism

4. Which of these solid figures can only be made from one type of polygon?

A. triangular prism

B. triangular pyramid

C. rectangular pyramid

D. rectangular prism

5. Which of the following is true about a cube?

I It has 8 vertices.

II It has 6 faces.

III It has 8 edges.

A. I and II

B. II and III

C. I and III

D. I, II, and III

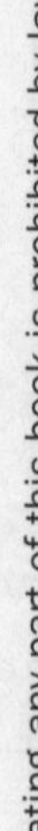

6. Which solid figure can be made from this net?

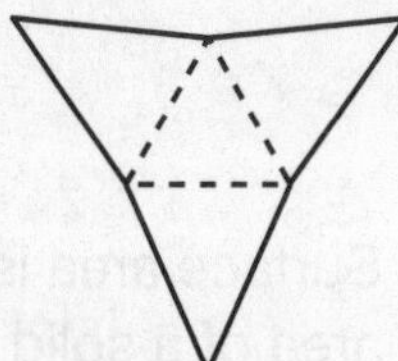

A. rectangular prism
B. rectangular pyramid
C. triangular pyramid
D. triangular prism

7. Which three-dimensional figure can be made from this net?

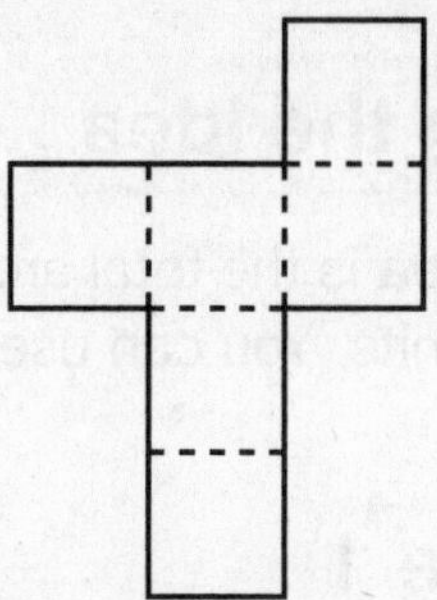

A. rectangular pyramid
B. cube
C. triangular prism
D. triangular pyramid

8. A triangular pyramid has 4 faces, 6 edges, and 4 vertices. A rectangular pyramid has 5 faces, 8 edges, and 5 vertices.

A. How many faces, edges, and vertices would a pentagonal pyramid have?

B. What kind or kinds of polygons would you need to construct a pentagonal pyramid? How many of each kind or kinds of polygons would you need?

Common Core State Standard:
6.G.4

Surface Area

Getting the Idea

Surface area is the total area of the surfaces of a solid figure. Surface area is measured in square units. You can use a net to help you find the surface area of a solid figure.

Example 1

What is the surface area of this rectangular prism?

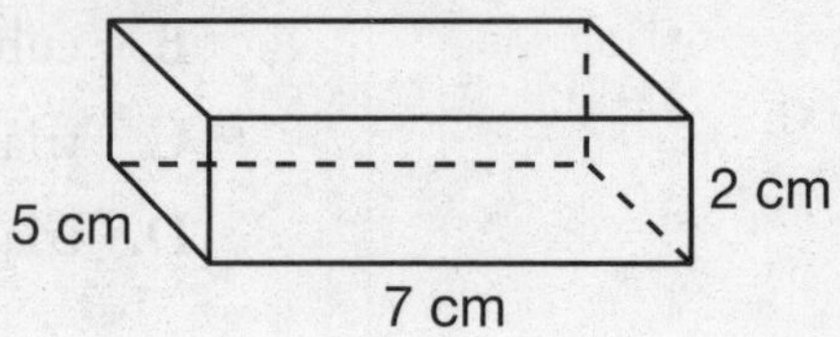

Strategy **Use a net.**

Step 1 Draw the net showing the dimensions.

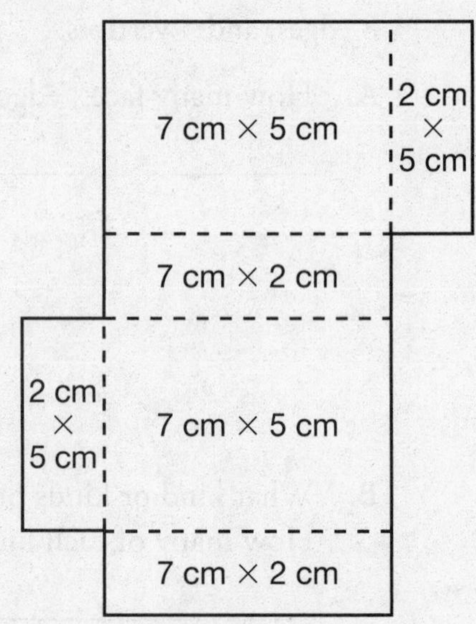

Find the area of each face.

$A = 7 \times 5 = 35$

$A = 2 \times 5 = 10$

$A = 7 \times 2 = 14$

$A = 2 \times 5 = 10$

$A = 7 \times 5 = 35$

$A = 7 \times 2 = 14$

Step 2 Add the areas to find the total surface area.

$35 + 10 + 14 + 10 + 35 + 14 = 118$

Solution **The surface area of the rectangular prism is 118 cm^2.**

A formula can be used to find the surface area (*SA*) of a rectangular prism:

$SA = 2lw + 2lh + 2wh$, where l is the length, w is the width, and h is the height

You may also see *S.A.* used for surface area in some formulas.

Example 2

What is the surface area of this rectangular prism?

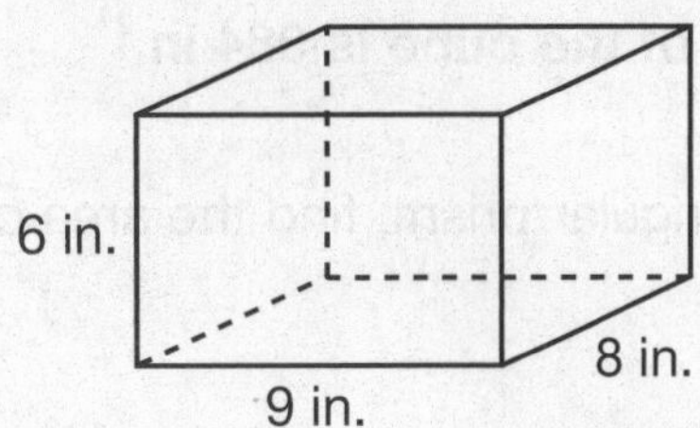

Strategy **Use the formula for the surface area of a rectangular prism.**

Step 1 Identify the dimensions of the prism.

$l = 9$ in. $w = 8$ in. $h = 6$ in.

Step 2 Substitute the values into the formula.

$A = 2lw + 2lh + 2wh$

$A = (2 \times 9 \times 8) + (2 \times 9 \times 6) + (2 \times 8 \times 6)$

Step 3 Multiply.

$A = (2 \times 9 \times 8) + (2 \times 9 \times 6) + (2 \times 8 \times 6)$

$A = (18 \times 8) + (18 \times 6) + (16 \times 6) = 144 + 108 + 96$

Step 4 Add.

$A = 144 + 108 + 96 = 348$

Solution **The surface area of the rectangular prism is 348 in.^{2}**

The formula for the surface area of a cube is $SA = 6s^2$, where s is the length of a side of the cube.

Example 3

What is the surface area of this cube?

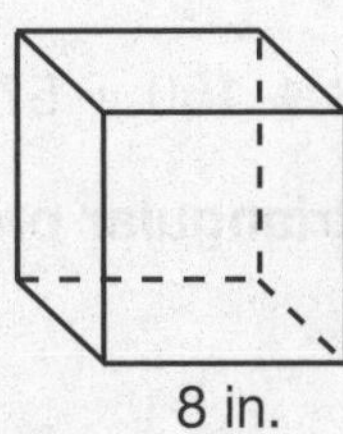

Strategy **Use the formula for the surface area of a cube.**

Step 1 Substitute the values into the formula.

$SA = 6s^2 = 6 \times 8^2$

Step 2 Multiply.

$SA = 6 \times 8^2 = 6 \times 8 \times 8 = 384$

Solution **The surface area of the cube is 384 in.^{2}**

To find the surface area of a triangular prism, find the area of each of the faces. Then add the areas. Use a net to help you.

Example 4

What is the surface area of this triangular prism?

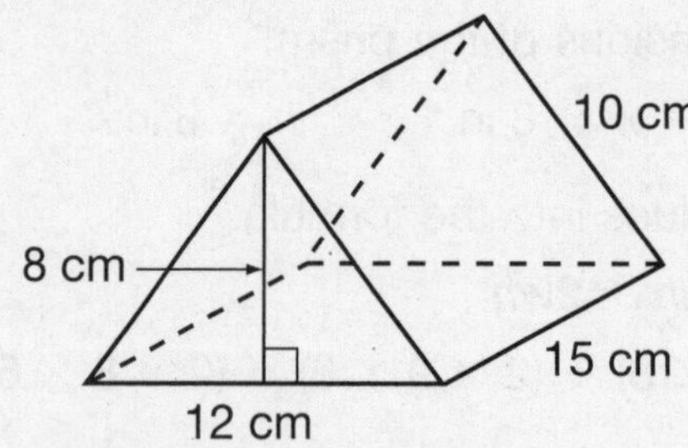

Strategy **Use a net. Add the areas of the faces.**

Step 1 Draw the net showing the dimensions.

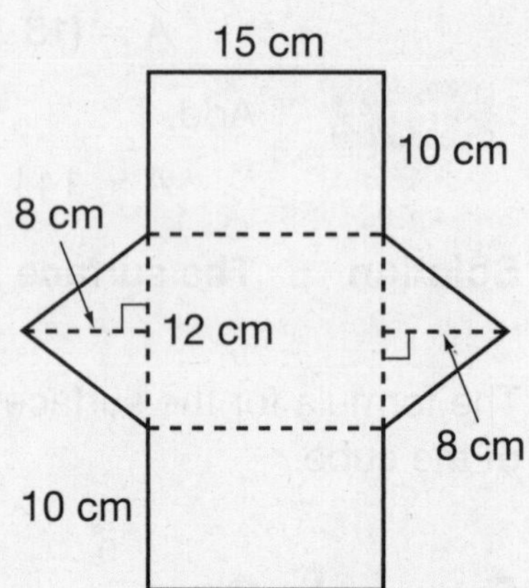

Find the area of each face.

Use $A = \frac{1}{2}bh$ for the triangles and $A = lw$ for the rectangles.

$A = 15 \times 10 = 150$

$A = \frac{1}{2} \times 12 \times 8 = 48$

$A = 15 \times 12 = 180$

$A = \frac{1}{2} \times 12 \times 8 = 48$

$A = 15 \times 10 = 150$

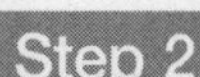

Step 2 Add the areas.

$150 + 48 + 180 + 48 + 150 = 576$

Solution **The surface area of the triangular prism is 576 cm^2.**

Remember that a pyramid has one base, which can be any polygon. All of the other faces of the pyramid, known as **lateral faces**, are triangles.

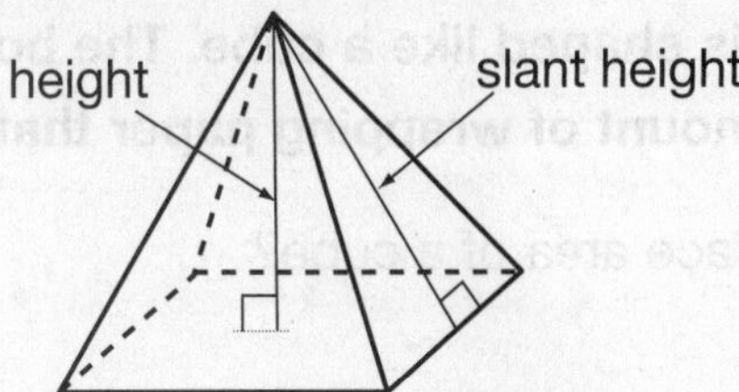

The surface area of a pyramid is the sum of the areas of its base and lateral faces. To find the area of a lateral face, you can use the formula for the area of a triangle: $A = \frac{1}{2}bh$. The only difference is that you should use the **slant height** of the lateral face. It is important that you do not use the height of the pyramid in this formula.

Example 5

What is the surface area of this square pyramid?

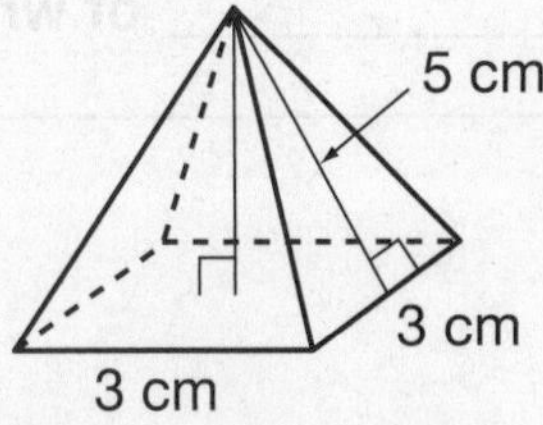

Strategy **Use a net. Add the areas of the faces.**

Step 1 Draw the net showing the dimensions.

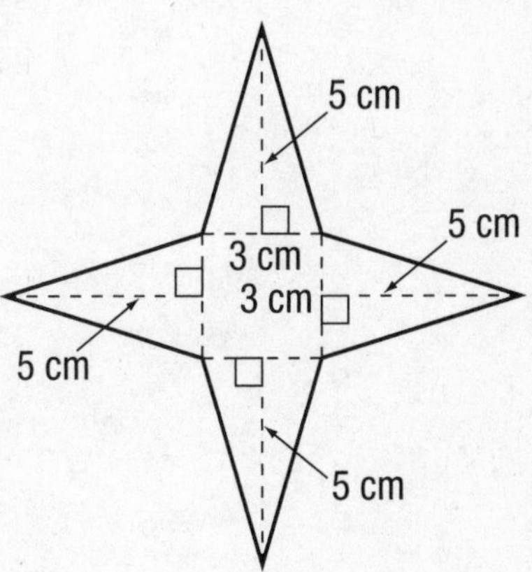

Step 2 Find the area of the square base.

$A = s^2 = 3 \times 3 = 9$

Step 3 Find the total area of the four triangular faces.

Each triangular face has the same base and slant height.

Find the area of one triangular face.

$A = \frac{1}{2}bh = \frac{1}{2} \times 3 \times 5 = 7.5$

Multiply the area by 4.

$4 \times 7.5 = 30$

Step 4 Add the areas of base and the faces.

$9 + 30 = 39$

Solution **The surface area of the pyramid is 39 cm^2.**

Coached Example

Akira is wrapping a box that is shaped like a cube. The box has a side length of 5 inches. What is the least amount of wrapping paper that Akira needs?

What is the formula for the surface area of a cube?

$SA =$ __________

Substitute the known value into the formula.

$SA =$ ______________

What is the surface area of the box? ______________

Akira will need at least ______________________ of wrapping paper to wrap the box.

Lesson Practice

Choose the correct answer.

1. The net for a rectangular prism is shown below.

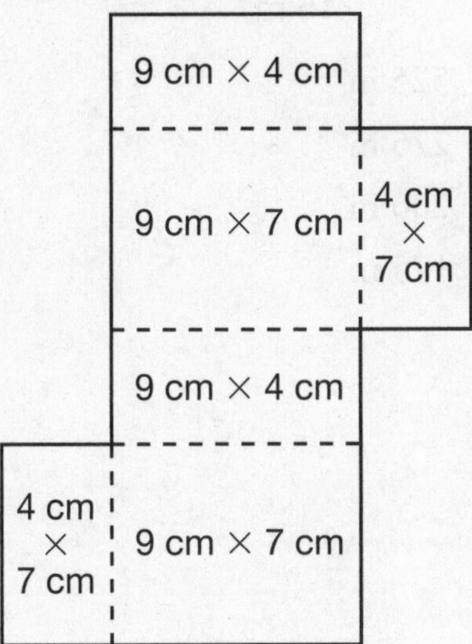

What is the surface area of the rectangular prism?

A. 504 cm^2

B. 254 cm^2

C. 252 cm^2

D. 127 cm^2

2. What is the surface area of a cube with edge lengths of 12 inches?

A. 144 $in.^2$

B. 432 $in.^2$

C. 864 $in.^2$

D. 1,728 $in.^2$

3. What is the surface area of this rectangular prism?

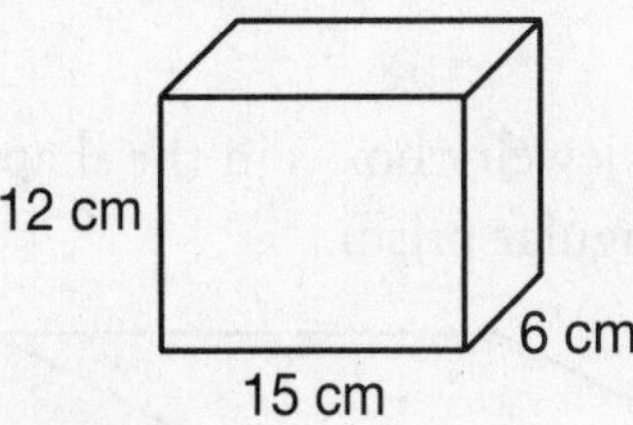

A. 342 cm^2

B. 540 cm^2

C. 684 cm^2

D. 1,080 cm^2

4. What is the surface area of this triangular prism?

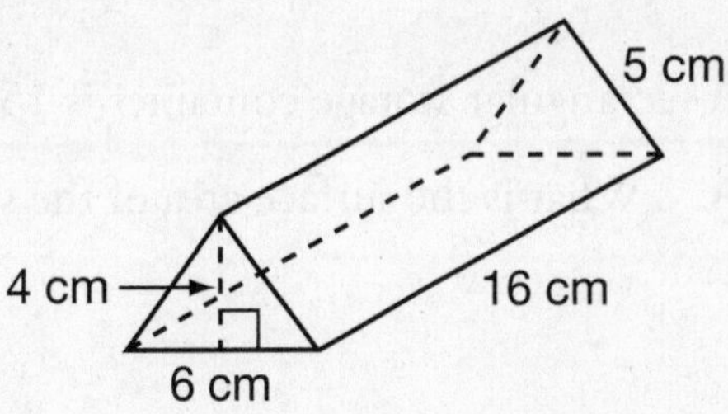

A. 264 cm^2

B. 280 cm^2

C. 312 cm^2

D. 324 cm^2

5. Helena wants to paint a box in the shape of a cube with sides that are 18 inches long. What is the surface area that Helena will paint?

A. 324 in.^{2} C. 1,296 in.^{2}

B. 648 in.^{2} D. 1,944 in.^{2}

6. Erin's jewelry box is in the shape of a rectangular prism.

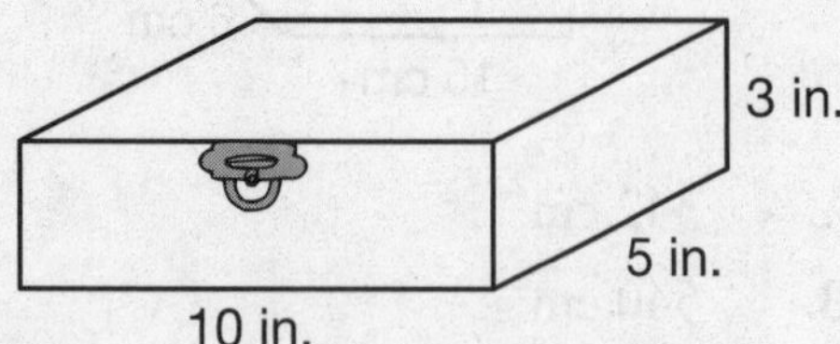

What is the surface area of Erin's jewelry box?

A. 95 in.^{2} C. 160 in.^{2}

B. 150 in.^{2} D. 190 in.^{2}

7. What is the surface area of the square pyramid.

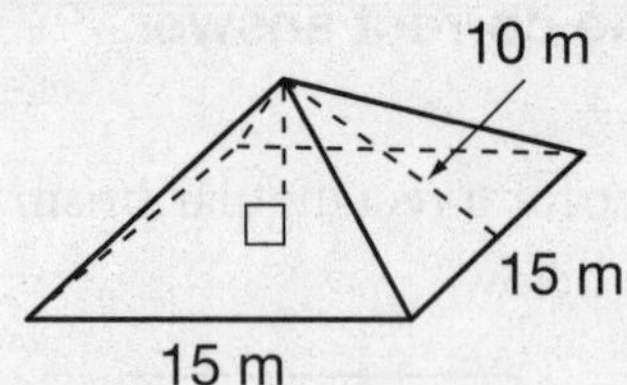

A. 525 m^2

B. 275 m^2

C. 200 m^2

D. 175 m^2

8. A rectangular storage container is 15 feet long, 12 feet wide, and 8 feet high.

A. What is the surface area of the storage container, including the floor? Show your work.

B. Explain how you found your answer.

Common Core State Standard:
6.G.2

Volume

Getting the Idea

Volume (V) is a measure of the number of **cubic units** that fit inside a solid figure. A cubic unit can be any unit such as a cubic inch (in.^{3}) or a cubic centimeter (cm^3), both shown below.

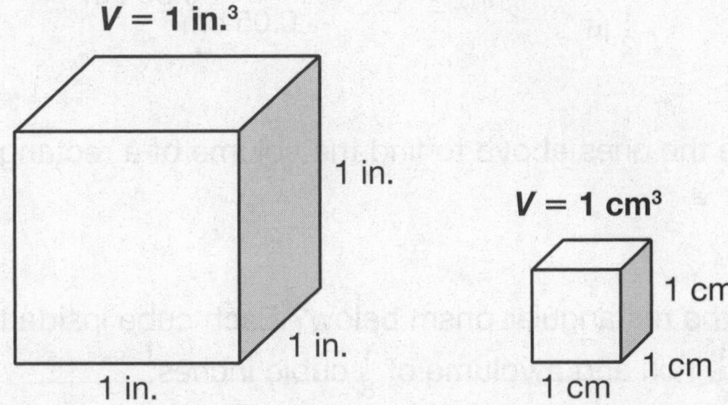

To find the volume of a rectangular prism or a cube, you can count the number of cubes that would fit inside the figure.

Example 1

What is the volume of the rectangular prism?

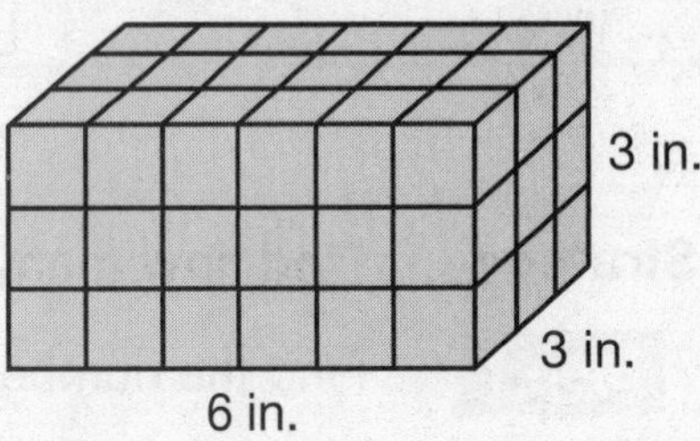

Strategy **Find how many cubes fit inside the prism.**

Find the number of cubes in the bottom layer.

There are 3 rows of 6 cubes each.

So, there are $3 \times 6 = 18$ cubes in the bottom layer.

Find the total number of cubes in the prism.

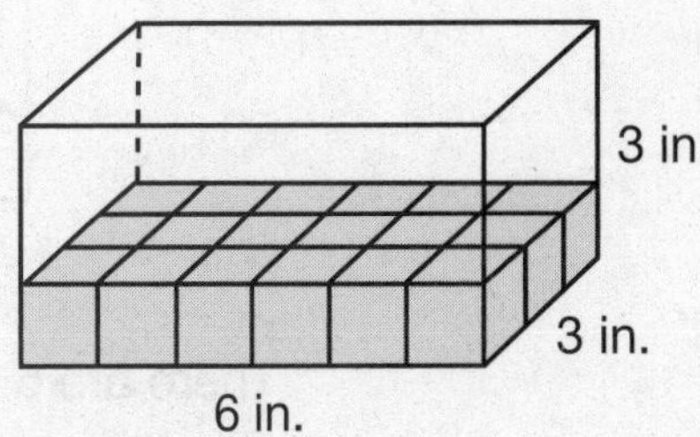

Multiply the number of cubes in the bottom layer by the number of layers.

$3 \times 18 = 54$

Each cube inside the rectangular prism measures 1 cubic inch.

The volume is 54 cubic inches.

Solution **The volume of the rectangular prism is 54 cubic inches.**

Sometimes the edge lengths of a prism include fractions or decimals. For example, a cube with a volume of $\frac{1}{8}$ cubic inch has an edge length of $\frac{1}{2}$ inch and a cube with a volume of 0.125 cubic centimeters has an edge length of 0.5 centimeter.

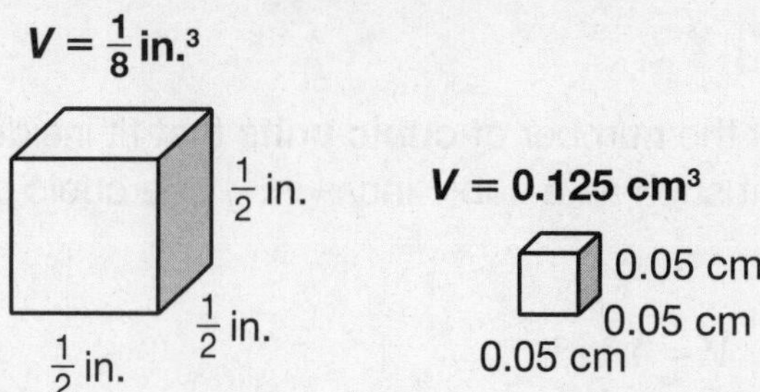

You can use cubes like the ones above to find the volume of a rectangular prism.

Example 2

What is the volume of the rectangular prism below? Each cube inside the rectangular prism has an edge length of $\frac{1}{2}$ inch and a volume of $\frac{1}{8}$ cubic inches.

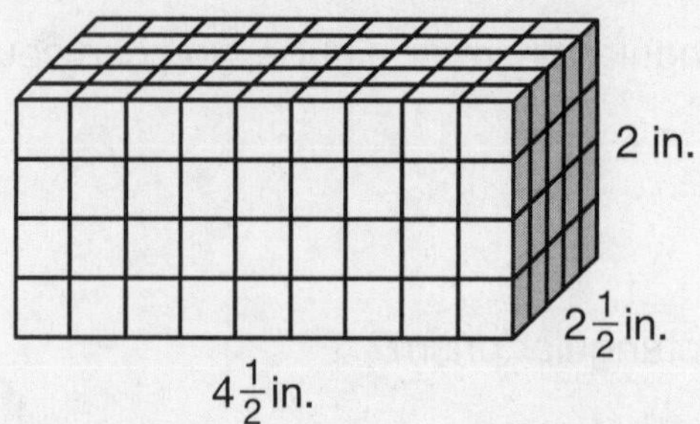

Strategy **Find how many cubes fit inside the prism. Multiply to find the volume.**

Step 1 Find the number of cubes in the bottom layer.

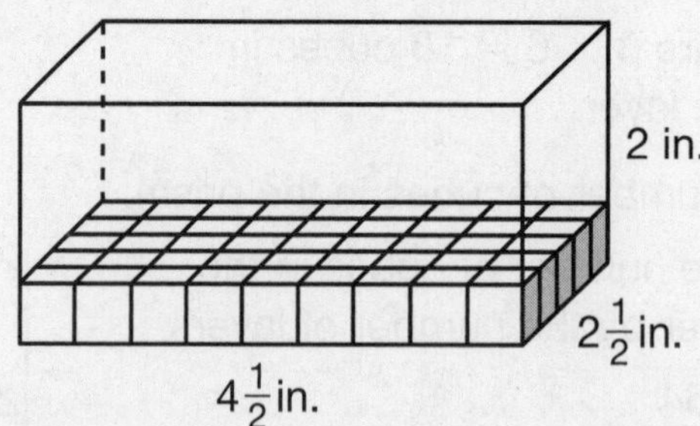

There are 5 rows of 9 cubes each.

So, there are $5 \times 9 = 45$ cubes in the bottom layer.

Step 2 Find the total number of cubes in the prism.

Multiply the number of cubes in the bottom layer by the number of layers.

There are 4 layers.

$4 \times 45 = 180$

There are 180 cubes, each with an edge length of $\frac{1}{2}$ inch, in the prism.

Step 3 Multiply to find the volume of the prism.

Each cube has a volume of $\frac{1}{8}$ in.^{3}, so the volume of the prism is:

$$180 \times \frac{1}{8} = \frac{\overset{45}{\cancel{180}}}{1} \times \frac{1}{\underset{2}{\cancel{8}}} = \frac{45}{2} = 22\frac{1}{2}$$

Solution **The volume of the rectangular prism is $22\frac{1}{2}$ in.^{3}**

Note: You could also have used multiplication to find the volumes of the prisms in Examples 1 and 2. You may have noticed that the volume of each prism is equal to length $\times$ width $\times$ height.

In Example 1: volume $= 6 \times 3 \times 3 = 54$ cubic inches

In Example 2: volume $= 4\frac{1}{2} \times 2\frac{1}{2} \times 2 = 22\frac{1}{2}$ cubic inches

You can use formulas to find the volumes of rectangular prisms or cubes.

Formula for Volume, V	Diagram
Rectangular Prism $V = lwh$, where l represents the length, w represents the width, and h represents the height. This formula is also written as $V = bh$, where b represents the area of the base (length $\times$ width) and h represents the height.	h w l
Cube $V = s^3$ or $V = s \times s \times s$, where s represents the side length.	s s s

Example 3

What is the volume of this fish tank?

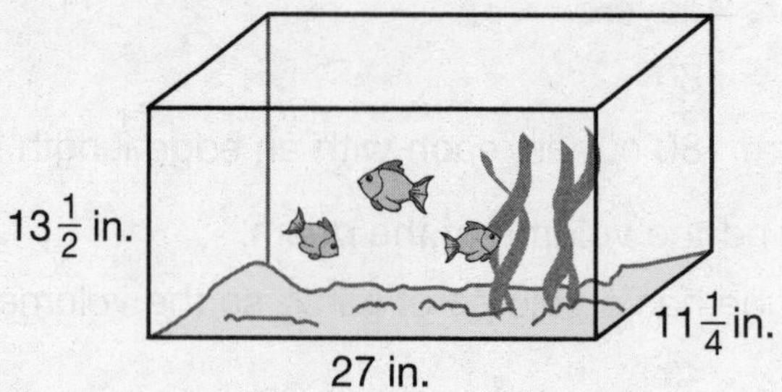

Strategy **Use the formula for the volume of a rectangular prism.**

Step 1 Substitute known values into the formula, $V = lwh$.

Length = 27 in., width = $11\frac{1}{4}$ in., and height = $13\frac{1}{2}$ in.

$V = lwh$

$V = 27 \times 11\frac{1}{4} \times 13\frac{1}{2}$

Step 2 Solve.

$V = 27 \times 11\frac{1}{4} \times 13\frac{1}{2}$

$V = \frac{27}{1} \times \frac{45}{4} \times \frac{27}{2} = \frac{32,805}{8} = 4,100\frac{5}{8}$

Solution **The volume of the fish tank is $4,100\frac{5}{8}$ cubic inches, or $4,100\frac{5}{8}$ in.^{3}**

Note: You can also use the formula $V = Bh$ to find the volume of the fish tank in Example 3. Remember that in $V = Bh$, B represents the area of the base, and that the area of a rectangle equals length × width, or lw.

$$V = B \times h$$
$$= \left(27 \times 11\frac{1}{4}\right) \times 13\frac{1}{2}$$
$$= \left(\frac{27}{1} \times \frac{45}{4}\right) \times \frac{27}{2}$$
$$= \frac{1,215}{4} \times \frac{27}{2}$$
$$= \frac{32,805}{8} = 4,100\frac{5}{8}$$

Example 4

What is the volume of this cube?

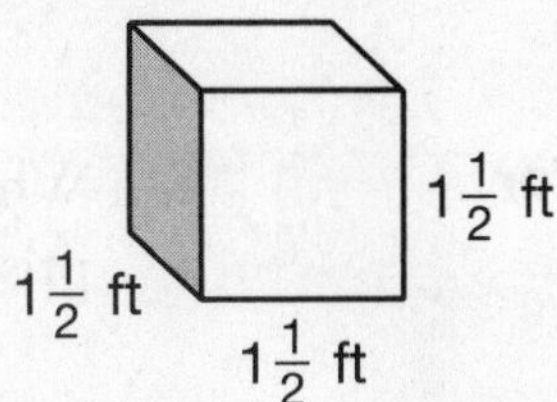

Strategy **Use the formula for the volume of a cube.**

Each side, s, measures $1\frac{1}{2}$ feet or $\frac{3}{2}$ feet.

$$V = s^3 = s \times s \times s$$

$$V = \frac{3}{2} \times \frac{3}{2} \times \frac{3}{2} = \frac{27}{8} = 3\frac{3}{8}$$

Solution **The volume of the cube is $3\frac{3}{8}$ cubic feet, or $3\frac{3}{8}$ ft^3.**

Coached Example

What is the volume of a rectangular prism with the dimensions shown below?

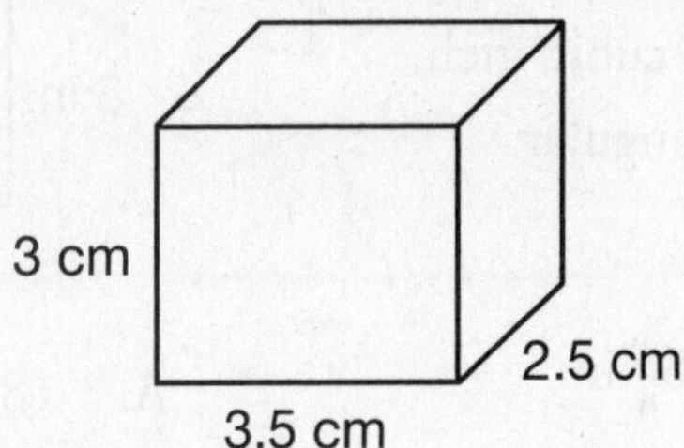

The formula for the volume of a rectangular prism is $V =$ __________.

Substitute known values into the formula.

Length (l) = __________ Width (w) = __________ Height (h) = __________

$V =$ __________ × __________ × __________

$V =$ __________

The volume of the prism shown above is __________ cubic centimeters.

Lesson Practice

Choose the correct answer.

Use the rectangular prism below for questions 1 and 2.

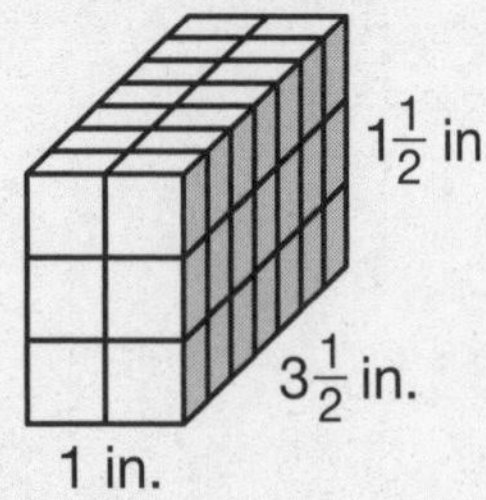

1. How many cubes are inside the rectangular prism?

   A. 27 cubes **C.** 41 cubes
   B. 28 cubes **D.** 42 cubes

2. If each cube has a volume of $\frac{1}{8}$ cubic inch, what is the volume of the rectangular prism?

   A. $3\frac{1}{4}$ in.^{3} **C.** $6\frac{1}{4}$ in.^{3}
   B. $5\frac{1}{4}$ in.^{3} **D.** 42 in.^{3}

3. What is the volume of the cube shown below?

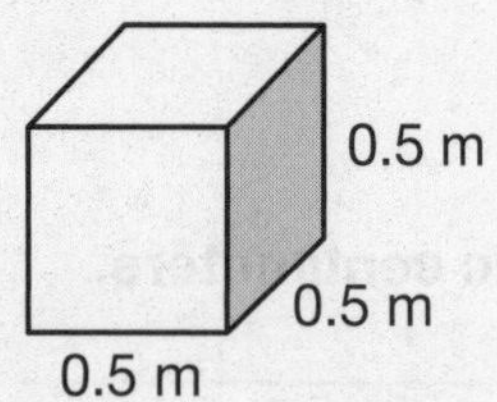

   A. 0.125 m^3 **C.** 1.25 m^3
   B. 0.25 m^3 **D.** 1.5 m^3

4. What is the volume of this rectangular prism?

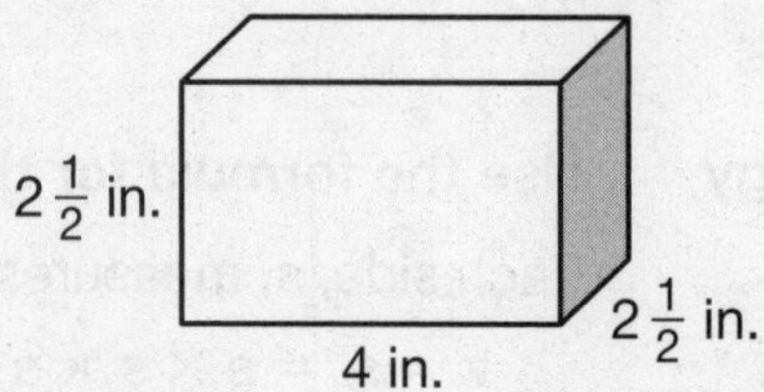

   A. 9 in.^{3} **C.** 50 in.^{3}
   B. 25 in.^{3} **D.** 100 in.^{3}

5. What is the volume of this rectangular prism?

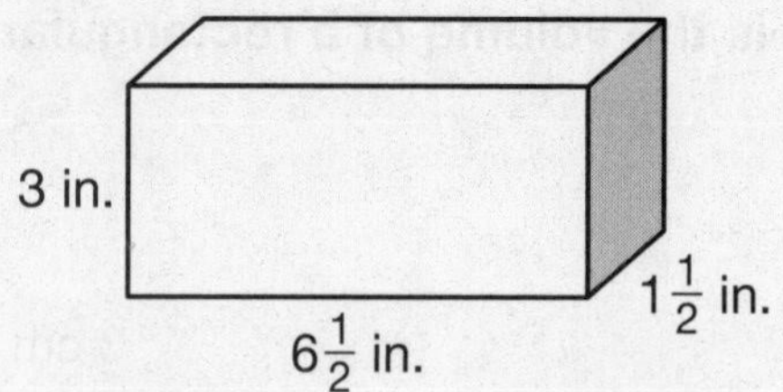

   A. $18\frac{3}{8}$ in.^{3} **C.** $29\frac{1}{4}$ in.^{3}
   B. 27 in.^{3} **D.** $40\frac{1}{2}$ in.^{3}

6. What is the volume of this cube?

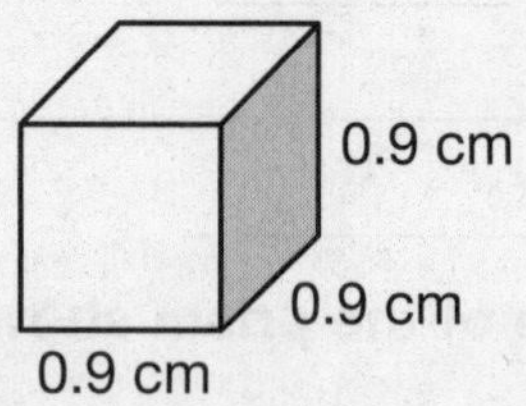

   A. 7.29 cm^3
   B. 0.729 cm^3
   C. 0.27 cm^3
   D. 0.027 cm^3

7. Keiko bought this plastic storage box for her room. What is the volume of the box?

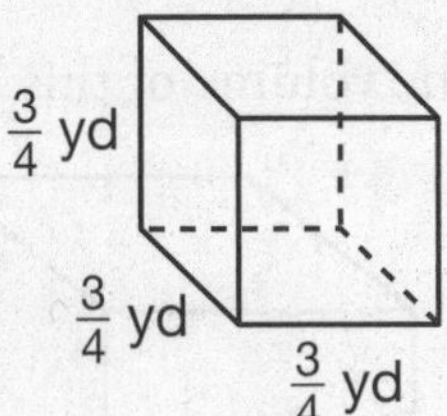

A. $\frac{27}{64}$ yd^3

B. $\frac{3}{4}$ yd^3

C. $2\frac{1}{4}$ yd^3

D. $6\frac{3}{4}$ yd^3

8. What is the maximum number of cubic feet of water this swimming pool can hold?

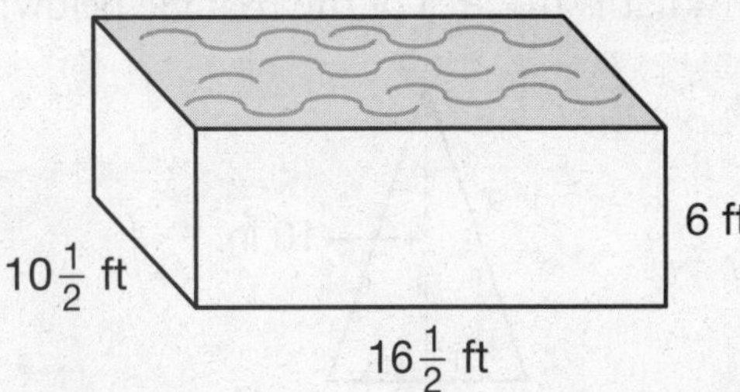

A. $1{,}039\frac{1}{2}$ ft^3

B. $1{,}030\frac{1}{2}$ ft^3

C. $961\frac{1}{2}$ ft^3

D. $960\frac{1}{4}$ ft^3

9. A self-storage facility sells the two boxes shown below.

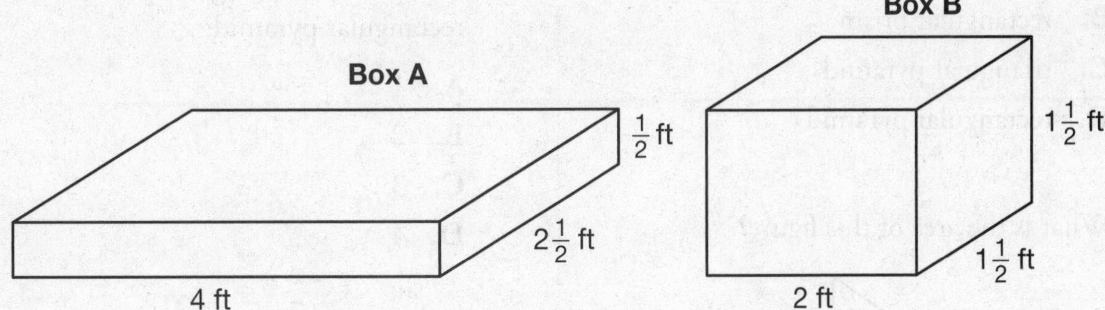

A. Find the volume of Box *A*, in cubic feet, showing each step in the process.

____

B. Which box has the greater volume, box *A* or box *B*? Show each step of your work.

____

Domain 4: Cumulative Assessment for Lessons 24–31

1. What is the area of the triangle below?

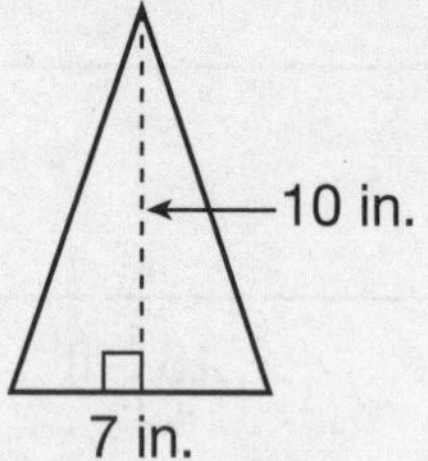

A. 70 square inches
B. 35 square inches
C. 25 square inches
D. 17.5 square inches

2. Which solid figure has 4 faces that are triangles and 1 face that is a rectangle?

A. triangular prism
B. rectangular prism
C. triangular pyramid
D. rectangular pyramid

3. What is the area of this figure?

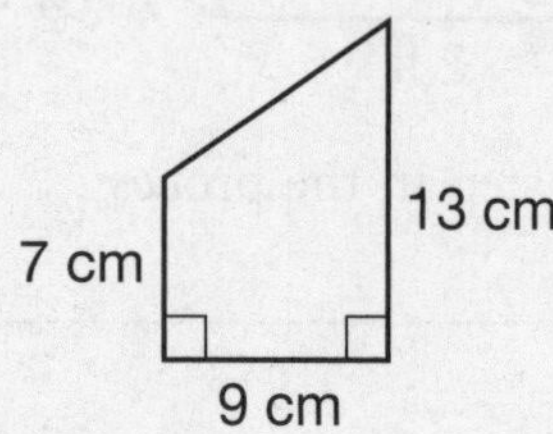

A. 63 cm^2
B. 90 cm^2
C. 91 cm^2
D. 117 cm^2

4. What is the volume of this box?

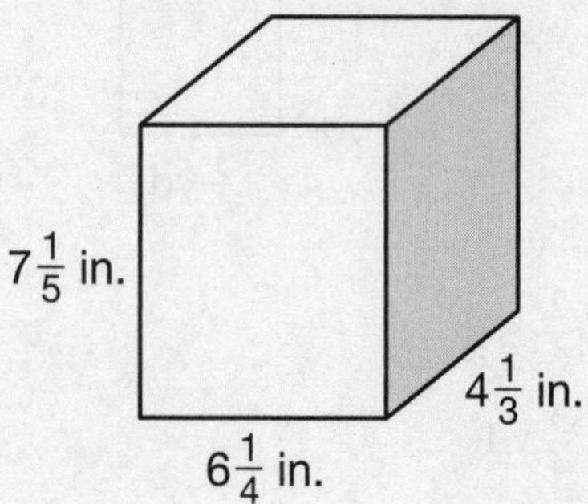

A. 168 $in.^3$
B. 195 $in.^3$
C. 224 $in.^3$
D. 585 $in.^3$

5. What is the difference in the number of edges in a rectangular prism and a rectangular pyramid?

A. 1
B. 2
C. 3
D. 4

6. Keeshawn measured the box shown below.

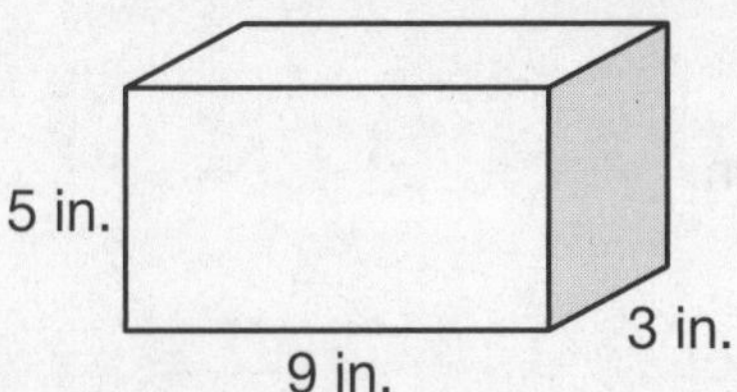

What is the surface area of the box?

A. 175 in.^{2}

B. 174 in.^{2}

C. 135 in.^{2}

D. 87 in.^{2}

7. A rectangle is drawn on the coordinate plane below.

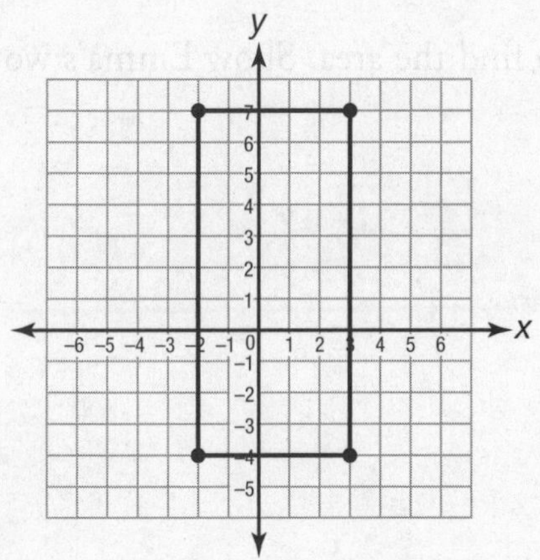

What is the perimeter of the rectangle?

A. 11 units

B. 22 units

C. 32 units

D. 55 units

8. A storage room measures $9\frac{1}{2}$ feet by 6 feet by $9\frac{1}{2}$ feet. What is the volume of the storage room?

A. 57 cubic feet

B. 486 cubic feet

C. $541\frac{1}{2}$ cubic feet

D. $586\frac{5}{8}$ cubic feet

9. The net of a square pyramid is shown below.

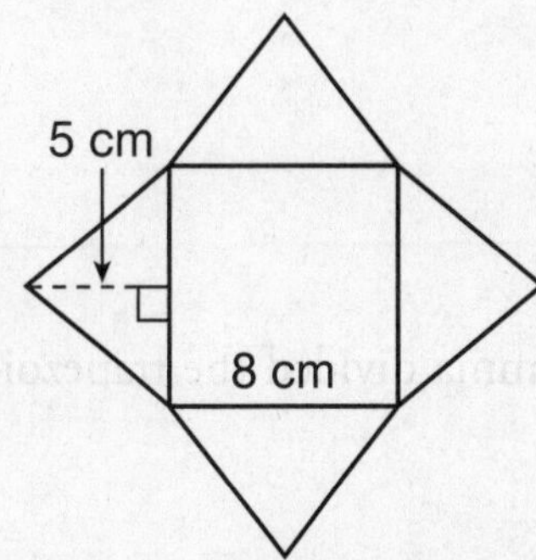

What is the surface area of the pyramid?

____

10. Jonathan and Emma found the area of the trapezoid below.

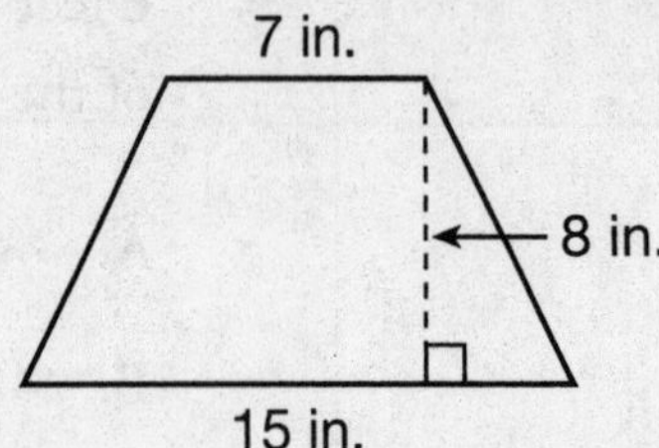

A. Jonathan used the formula for the area of a trapezoid to find the area. Show Jonathan's work.

B. Emma divided the trapezoid into two triangles to find the area. Show Emma's work.

C. What is the area of the trapezoid?

____

Domain 5

Statistics and Probability

Domain 5: Diagnostic Assessment for Lessons 32–37

1. Which of the following questions is a statistical question?

A. How old is the school's principal?

B. How tall is each student in your class?

C. How many students are in the school band?

D. What is my favorite number?

2. Jeanne's math test scores were 80, 95, 90, 80, and 100. What is her mean score?

A. 80

B. 89

C. 90

D. 91

3. What is the interquartile range for the data on the box plot?

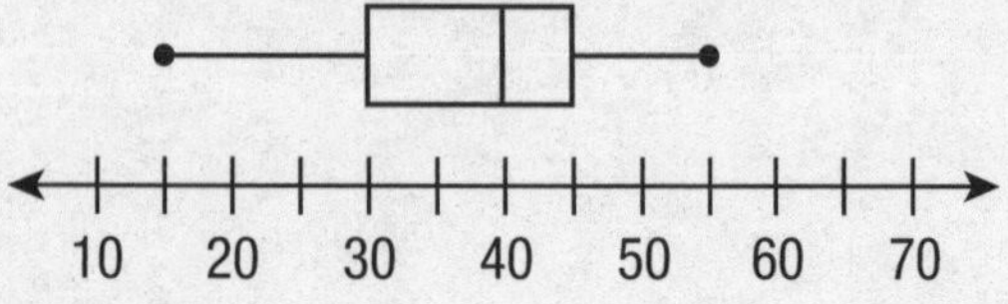

A. 15

B. 30

C. 40

D. 55

Use the histogram for questions 4 and 5.

The histogram shows the number of points that the Cougars basketball team scored in each game this season.

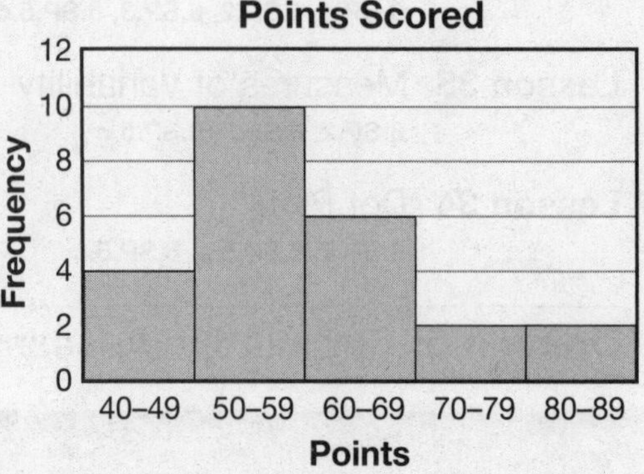

4. In how many games did the Cougars score from 50 to 79 points?

A. 10

B. 16

C. 18

D. 20

5. The Cougars won every time they scored at least 60 points. They won half of their other games. How many games did the Cougars win?

A. 10

B. 14

C. 15

D. 17

Use the dot plot for questions 6 and 7.

The dot plot shows how many hours each student in a class worked on a research project.

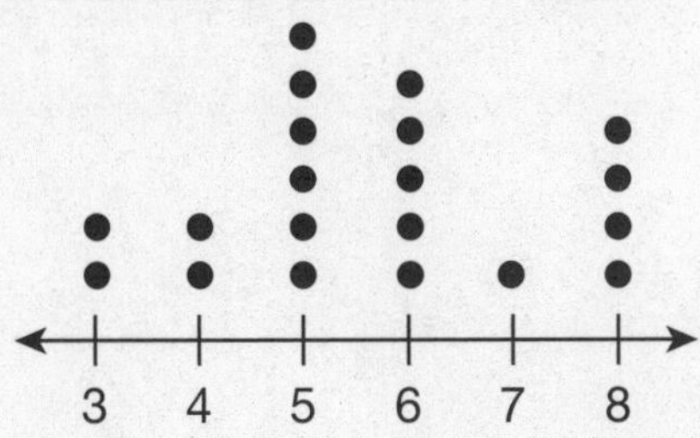

6. How many students spent more than 6 hours working on the project?

   A. 1
   B. 3
   C. 5
   D. 10

7. Which statement is **not** true?

   A. The data clusters around 5 and 6 hours.
   B. The mode is 5 hours.
   C. Twenty classmates worked on the class project.
   D. There is a gap in the data.

8. For which set of data would the median be best to describe the data?

   A. the selling price of a gallon of gas
   B. the ages of all the people in your family including your grandparents, aunts, uncles, and cousins
   C. the sizes of suits sold in a men's store
   D. the sandwiches sold for lunch in one day

9. The following data set shows the number of minutes it took students to complete an obstacle course.

   9, 7, 10, 11, 13, 12,
   8, 10, 12, 11, 10, 9

Make a dot plot of the data.

Minutes to Complete Obstacle Course

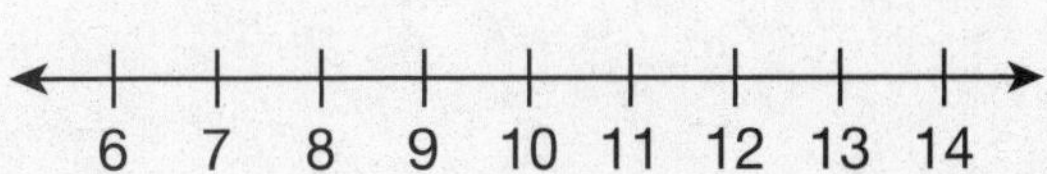

10. The heights of the players on a basketball team, in inches, are shown below.

74, 81, 77, 75, 73

A. Find the interquartile range of the data set. Show your work.

____

B. How does the interquartile range reflect the heights of the basketball players?

____

____

____

Measures of Center

Common Core State Standards:
6.SP.1, 6.SP.2, 6.SP.3, 6.SP.5.c

Getting the Idea

Not all questions are statistical questions. For example, if you ask 1 friend "How much time did you spend watching TV yesterday?" you will get just 1 answer. But if you ask 50 people the same question, the answers you get will vary. A statistical question is any question for which you expect to get a variety of answers.

Example 1

Which of the following questions is a statistical question?

1. How tall is the town's mayor?
2. What are the heights of the players on the school basketball team?

Strategy **Determine whether answers to the question will vary.**

Step 1 Analyze question 1.

The question is "How tall is the town's mayor?"

A town has only one mayor. There is only one answer to the question: the height of the person who is mayor.

Step 2 Analyze question 2.

The question is "What are the heights of the players on the school basketball team?"

There are many players on the basketball team. Those players are probably not all the same height.

You expect to get many different answers to this question.

Solution **Question 1 is not a statistical question. Question 2 is a statistical question.**

A **measure of center** is a single number that you can use to describe all of the values in a data set. You can think of a measure of center as a number that tells you roughly what the middle or average value in a data set is. Mean, median, and mode are all measures of center.

The **mean** is equal to the sum of the terms in a data set divided by the number of terms in the set.

The **median** is the middle term in a data set ordered from least to greatest. If there is an even number of terms in a set, the median is the mean of the two middle numbers.

The **mode** is the term or terms that appear most frequently in the data set. A set may have no mode, one mode, or more than one mode.

Example 2

Frank asked 5 of his relatives how many miles they drive to work. Their answers are 20 miles, 24 miles, 12 miles, 3 miles, and 16 miles. Find the mean and the median of the data.

Strategy **Use the definitions of mean and median.**

Step 1 Find the mean of the data.

The mean is the quotient of the sum of the terms divided by the number of terms.

Add the terms.

$3 + 12 + 16 + 20 + 24 = 75$

Divide that sum by the number of terms. There are 5 terms.

$75 \div 5 = 15$

The mean is 15 miles.

Step 2 Find the median of the data.

The median is the middle term in a data set ordered from least to greatest.

Order the data from least to greatest.

3, 12, 16, 20, 24

The middle term is 16.

The median is 16 miles.

Solution **The mean is 15 miles and the median is 16 miles.**

Notice that both 15 miles and 16 miles are roughly in the middle of the values in the data set. You could use either the mean or the median as a good way to describe the typical distance driven to work by Frank's relatives.

Example 3

The list below shows the weights, in pounds, of six dogs at a dog show:

52, 52, 55, 55, 54, 47

What are the median and the mode of the data?

Strategy **Use the definitions of median and mode.**

Step 1 Order the weights from least to greatest.

47, 52, 52, 54, 55, 55

Step 2 Find the median.

There is an even number of terms, so underline the two middle numbers.

47, 52, <u>52</u>, <u>54</u>, 55, 55

The median is the mean of those two middle numbers.

$52 + 54 = 106$ and $106 \div 2 = 53$

The median weight is 53 pounds.

Step 3 Find the mode.

The mode is the number or numbers that appear most often in a data set.

Two numbers appear twice in the data set: 52 and 55.

This data set has two modes: 52 pounds and 55 pounds.

Solution **The median weight is 53 pounds. There are 2 mode weights: 52 pounds and 55 pounds.**

Notice that the mode of 52 pounds seems to be a better value to describe the center of the weight data than the mode of 55 pounds. That is especially true if you look at it with the median weight of 53 pounds.

Example 4

The table shows Javier's scores for the last 4 games he bowled.

Javier's Bowling Scores

Game	1	2	3	4	5
Score	90	93	101	104	?

What score does Javier need in his fifth game to have a mean bowling score of exactly 100 for all 5 games?

Strategy **Use what you know about the mean to solve the problem.**

Step 1 Find the sum of the scores of the first 4 games.

$90 + 93 + 101 + 104 = 388$

Step 2 Find the total score Javier needs to have a mean score of 100 for all 5 games.

What number divided by 5 equals 100?

Multiply: $5 \times 100 = 500$

Javier needs a total score of 500.

Step 3 Find the number you need to add to 388 to get a sum of 500.

$500 - 388 = 112$

Javier needs to bowl 112 in game 5.

Step 4 Check your answer.

Find the mean if Javier's score for game 5 is 112.

Add: $90 + 93 + 101 + 104 + 112 = 500$

Divide: $500 \div 5 = 100$ ✓

Solution **Javier needs to score 112 points in game 5 to have a mean score of exactly 100 for all 5 games.**

If one data value is much greater or much less than the rest of the data values, it is called an **outlier**. An outlier may affect a measure of center.

Example 5

The low temperatures for five days last week were 20°F, 24°F, 18°F, 5°F, and 23°F. Which better describes the data, the median or the mean?

Strategy **Determine how the outlier affects the median and mean.**

Step 1 Identify the outlier.

The outlier is 5°F.

Step 2 Find the median.

Order the data from least to greatest: 5, 18, 20, 23, 24

The median is 20.

Step 3 Determine how the outlier affects the median.

Since the median is the middle number, it is not affected by the outlier.

Step 4 Find the mean.

Add: $20 + 24 + 18 + 5 + 23 = 90$

Divide: $90 \div 5 = 18$

The mean is 18.

Step 5 Determine how the outlier affects the mean.

Since the outlier is less than the mean, it lowers the mean.

Step 6 Check your answer. Find the mean of the data without the outlier.

Add: $20 + 24 + 18 + 23 = 85$

Divide: $85 \div 4 = 21.25$

Without the outlier, the mean is 21.25.

Solution **The median describes the data better than the mean because the outlier lowers the mean.**

In general, an outlier will affect the mean more than it will affect the median. If you have a data set with an outlier, the median usually will be a better choice as a measure of center.

Coached Example

Edie has taken 5 math quizzes. Her scores are: 89, 96, 92, 84, 94. What are Edie's mean and median scores?

The mean is equal to the ________ of the data items divided by the number of ________ ________.

Find the sum of Edie's scores.

_____ + _____ + _____ + _____ + _____ = _____

Divide the sum by _____ to find the mean score.

_____ ÷ _____ = _____

The median is the __________ value in a data set ordered from ________ to ____________________.

Order Edie's scores from least to greatest.

_____, _____, _____, _____, _____

What is the middle value? ________

Edie's mean score is ________. Edie's median score is ________.

Lesson Practice

Choose the correct answer.

1. Roger bowled 7 games last weekend. His scores are: 155, 165, 138, 172, 127, 193, 142. What is Roger's median score?

   A. 127 **C.** 156
   B. 155 **D.** 193

2. What is the mode of this data set?

   84, 92, 68, 79, 94,
   84, 92, 79, 84, 68

   A. 68 **C.** 84
   B. 79 **D.** 92

Use the data below for questions 3 and 4.

The number of pages that Carolyn wrote in her journal each day from Monday to Friday is shown below.

9, 8, 12, 6, 10

3. What is the mean number of pages she wrote per day?

   A. 5 **C.** 9
   B. 6 **D.** 11

4. What is the median number of pages she wrote per day?

   A. 5 **C.** 9
   B. 6 **D.** 11

Use the data below for questions 5–7.

The number of miles that Jenna cycled each week for a 7-week period is shown below.

36, 42, 28, 52, 48, 36, 31

5. What is the median number of miles Jenna cycled?

   A. 24
   B. 36
   C. 39
   D. 52

6. What is the mode number of miles Jenna cycled?

   A. 7
   B. 36
   C. 42
   D. There is no mode.

7. What is the mean number of miles Jenna cycled?

   A. 31
   B. 36
   C. 39
   D. 41

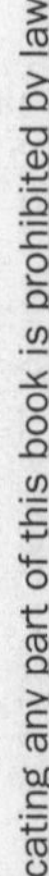

8. The number of wins that a college basketball team had each of the last five seasons is shown in the table below.

Wins per Season

Season	1	2	3	4	5	6
Number of Wins	27	18	24	25	12	?

What number of wins does the team need in season 6 to have a mean of 21 wins for all six seasons?

A. 18

B. 20

C. 22

D. 24

9. Joshua's math test scores were 81, 94, 90, 97, 50, and 86.

A. Find the median and the mean of the data.

median = ____________

mean = ____________

B. Which better describes the data, the median or the mean? Explain your answer.

____

____

____

Common Core State Standards: 6.SP.2, 6.SP.3, 6.SP.5.c

Measures of Variability

Getting the Idea

Like a measure of center, a **measure of variability** is a single number that can be used to describe an entire data set. The difference is that a measure of variability describes how spread out the data is, instead of describing the middle or the average of the data.

A common measure of the variability in a data set is the range. The **range** of a data set is the difference of the greatest value and the least value in the data set.

Example 1

Griffin's scores on his first seven math quizzes are shown below.

94, 86, 95, 86, 82, 90, 95

What is the range of Griffin's math quiz scores?

Strategy **Order the numbers from least to greatest. Then find the range.**

Step 1 Order the numbers from least to greatest.

82, 86, 86, 90, 94, 95, 95

Step 2 Identify the least and greatest values in the data.

The least value is 82.

The greatest value is 95.

Step 3 Find the difference of the greatest and least values.

$95 - 82 = 13$

Solution **The range of Griffin's quiz scores is 13.**

Remember that the median divides a data set into two halves. The median of the lower half of the data is called the **first quartile**. The median of the upper half of the data is called the **third quartile**. As shown in the diagram on the next page, the first quartile, median, and third quartile divide the full data set into four smaller data sets. The **interquartile range (IQR)** is the difference of the third quartile and the first quartile. Think of the IQR as the range of the middle half of the data. The IQR can be used to measure the variability of a data set.

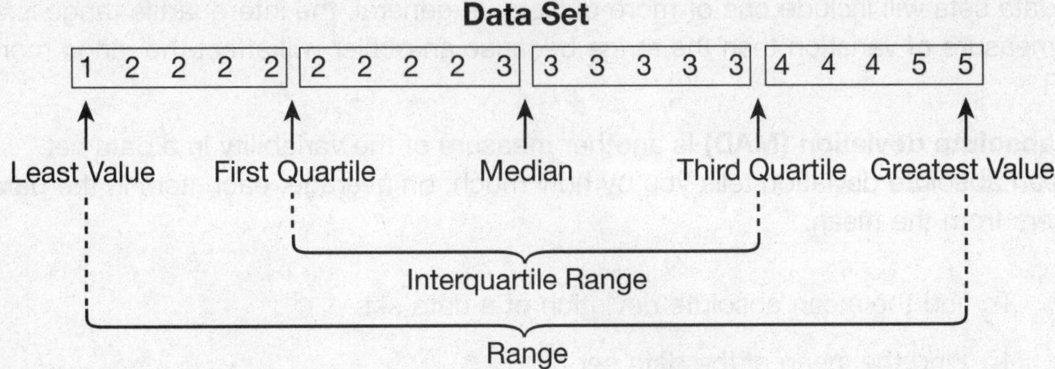

Example 2

The ages of Abe's grandchildren are: 20, 15, 23, 8, 20, 10, 15, 25, 16, and 18.

What are the first quartile, third quartile, and interquartile range of the grandchildren's ages?

Strategy **Order the numbers. Then find the quartiles.**

Step 1 Order the numbers from least to greatest.

8, 10, 15, 15, 16, 18, 20, 20, 23, 25

Step 2 Find the median.

There are 10 data items, so the median is the mean of the two middle numbers.

8, 10, 15, 15, <u>16</u>, <u>18</u>, 20, 20, 23, 25

$16 + 18 = 34$

$34 \div 2 = 17$

The median is 17.

Step 3 Find the first quartile.

The first quartile is the median of the lower half of the data:
8, 10, <u>15</u>, 15, 16.

The median of the lower half of the data is 15, so the first quartile is 15.

Step 4 Find the third quartile.

The third quartile is the median of the upper half of the data:
18, 20, <u>20</u>, 23, 25.

The median of the upper half of the data is 20, so the third quartile is 20.

Step 5 Find the interquartile range.

Subtract the first quartile from the third quartile: $20 - 15 = 5$

Solution **The first quartile is 15, the third quartile is 20, and the interquartile range is 5.**

Some data sets will include one or more outliers. In general, the interquartile range is a better measure of variation than the range because an outlier will affect the range more than the IQR.

Mean absolute deviation (MAD) is another measure of the variability in a data set. The mean absolute deviation tells you by how much, on average, each item in the data set differs from the mean.

To find the mean absolute deviation of a data set:

1. Find the mean of the data set.
2. Find the difference of each data item and the mean.
3. Find the absolute value of each difference, or deviation from the mean.
4. Find the sum of the absolute values.
5. Divide the sum by the total number of items in the data set.

Example 3

The ages of Abe's grandchildren are:

20, 15, 23, 8, 20, 10, 15, 25, 16, and 18.

What is the mean absolute deviation of the ages of Abe's grandchildren?

Strategy **Find the mean. Subtract each value from the mean. Then find the mean of the absolute values of the deviations from the mean.**

Step 1 Find the mean.

$20 + 15 + 23 + 8 + 20 + 10 + 15 + 25 + 16 + 18 = 170$

$170 \div 10 = 17$

Step 2 Find the absolute values of the deviations.

Make a table with the data items ordered from least to greatest.

Find the deviations from the mean. Then find their absolute values.

Use a calculator to find the deviation from the mean for each data item.

[8] [−] [1] [7] [=] [−9]

Data Item	Deviation from Mean	Absolute Value of Deviation from Mean
8	8 − 17 = −9	|−9| = 9
10	10 − 17 = −7	|−7| = 7
15	15 − 17 = −2	|−2| = 2
15	15 − 17 = −2	|−2| = 2
16	16 − 17 = −1	|−1| = 1
18	18 − 17 = 1	|1| = 1
20	20 − 17 = 3	|3| = 3
20	20 − 17 = 3	|3| = 3
23	23 − 17 = 6	|6| = 6
25	25 − 17 = 8	|8| = 8

Step 3 Find the mean absolute deviation.

Add the absolute values.

9 + 7 + 2 + 2 + 1 + 1 + 3 + 3 + 6 + 8 = 42

Divide by the number of items in the data set.

42 ÷ 10 = 4.2

The mean absolute deviation is 4.2.

Solution **The mean absolute deviation of the ages of Abe's grandchildren is 4.2.**

When a measure of variability for a data set is small, it means the data is clustered together. If the measure of variability is large, the data is more spread out.

Example 4

What do the interquartile range (IQR) and the mean absolute deviation (MAD) from Examples 1 and 2 tell you about the ages of Abe's grandchildren?

Strategy **Use the definitions of interquartile range and mean absolute deviation.**

Step 1 What does the IQR tell you about the ages of the grandchildren?

The IQR is 5.

That means that the middle half of the data values are within 5 of each other, from the ages of 15 to 20.

The IQR is small, so the ages of Abe's grandchildren are clustered together.

Step 2 What does the MAD tell you about the ages of the grandchildren?

The MAD is 4.2.

The MAD is small, so the ages of Abe's grandchildren are clustered together.

Solution **The interquartile range and the mean absolute deviation both show that the ages of Abe's grandchildren are close in value to each other.**

Coached Example

The number of text messages that Aimee sent on each of the past 7 days is shown below.

24, 19, 11, 15, 11, 28, 20

What are the first quartile, third quartile, and interquartile range of this data set?

Order the numbers from least to greatest.

______, ______, ______, ______, ______, ______, ______

Find the median. The median is the ____________ value of an ordered data set.

Median: ________

Find the first quartile. The first quartile is the median of the ____________ half of a data set.

First quartile: ________

Find the third quartile. The third quartile is the median of the ____________ half of a data set.

Third quartile: ________

The interquartile range is the ____________ of the __________ quartile and the __________ quartile.

Interquartile range: _____ − _____ = _____

Interquartile range: ________

The first quartile is ________, the third quartile is ________, and the interquartile range is ________.

Lesson Practice

Choose the correct answer.

Use the data below for questions 1–4.

33, 25, 42, 25, 31, 37, 46, 29, 38

1. What is the first quartile of the data?

A. 25

B. 27

C. 29

D. 33

2. What is the third quartile of the data?

A. 37

B. 38

C. 40

D. 46

3. What is the interquartile range of the data?

A. 8

B. 9

C. 13

D. 21

4. What is the mean absolute deviation of the data?

A. 6

B. 9

C. 21

D. 34

Use the data for questions 5–7.

The ages of the children at a birthday party are shown below.

7, 5, 10, 9, 15, 12, 6, 9, 7, 10

5. What is the first quartile of the data?

A. 5

B. 6

C. 7

D. 9

6. What is the third quartile of the data?

A. 9

B. 10

C. 11

D. 15

7. What is the interquartile range of the data?

A. 3

B. 5

C. 9

D. 10

8. Rosario recorded the weights of 2 litters of puppies 4 weeks after their births. The interquartile range for litter A was 3, and the interquartile range for litter B was 5. Which of the following statements must be true?

   A. The weights of litter A vary less than the weights of litter B.

   B. The weights of litter A vary more than the weights of litter B.

   C. The median weight of the puppies in litter A is less than that of litter B.

   D. The least weight of litter A is greater than the least weight of litter B.

9. Use the two data sets to answer the questions below.

   Data set A: 15, 23, 11, 6, 20

   Data set B: 18, 9, 14, 10, 15

   A. Find the interquartile range of each data set. Show your work.

   Data set A: ____________

   Data set B: ____________

   B. Which of the two data sets has less variability? Explain your answer.

   ____

   ____

   ____

Dot Plots

Getting the Idea

A **dot plot** uses a number line and dots to display numerical data. The number of dots above each value on the number line tells how many times that value occurs in a data set. Since a dot plot uses a number line, dot plots allow you to see the variation in a data set.

In a dot plot, a **cluster** shows where a group of data points fall. A **gap** is an interval where there are no data items.

Example 1

In a science class, the students weighed some samples to the nearest $\frac{1}{8}$ pound. The weights of the samples are given below.

$\frac{1}{8}$ lb, $\frac{3}{8}$ lb, $\frac{3}{4}$ lb, $\frac{1}{4}$ lb, $\frac{1}{8}$ lb, $\frac{1}{4}$ lb, $\frac{7}{8}$ lb, $\frac{1}{4}$ lb, $\frac{3}{8}$ lb, $\frac{1}{4}$ lb, $\frac{1}{2}$ lb, $\frac{3}{8}$ lb

Make a dot plot for the data. Then analyze the data.

Strategy **Create a dot plot to display the data.**

Step 1 Draw a number line.

The least value in the data is $\frac{1}{8}$ and the greatest value is $\frac{7}{8}$.

Draw a number line from 0 to 1, divided into eighths.

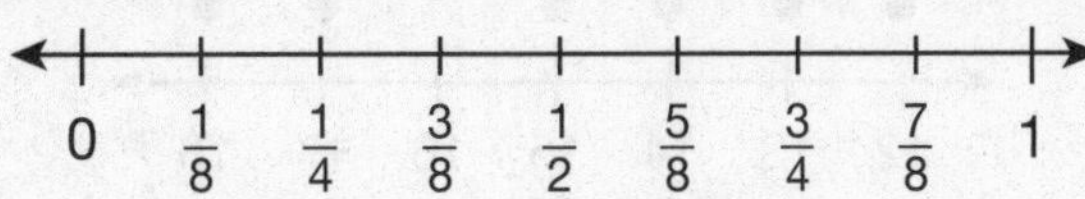

Step 2 Order the data values from least to greatest.

$\frac{1}{8}, \frac{1}{8}, \frac{1}{4}, \frac{1}{4}, \frac{1}{4}, \frac{1}{4}, \frac{3}{8}, \frac{3}{8}, \frac{3}{8}, \frac{1}{2}, \frac{3}{4}, \frac{7}{8}$

Step 3 Draw a dot for each value in the data set.

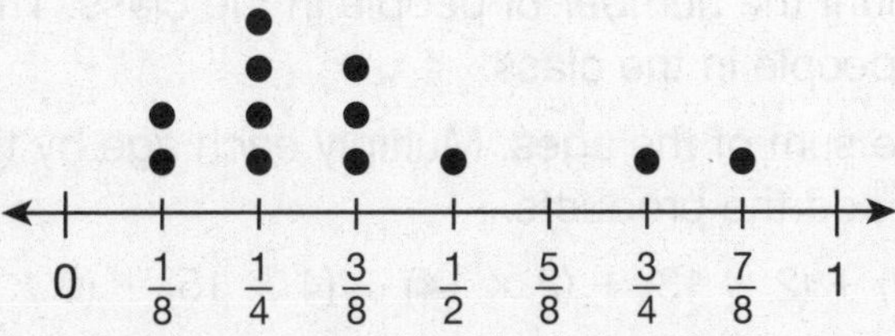

Step 4 Check your work.

First check that you have a dot for each sample weight.

There are 12 sample weights, so you should have 12 dots.

Then check that you drew the correct number of dots for each weight.

Step 5 Write a title for the dot plot.

The data is about the size of samples to the nearest $\frac{1}{8}$ pound.

Size of Samples to Nearest $\frac{1}{8}$ Pound

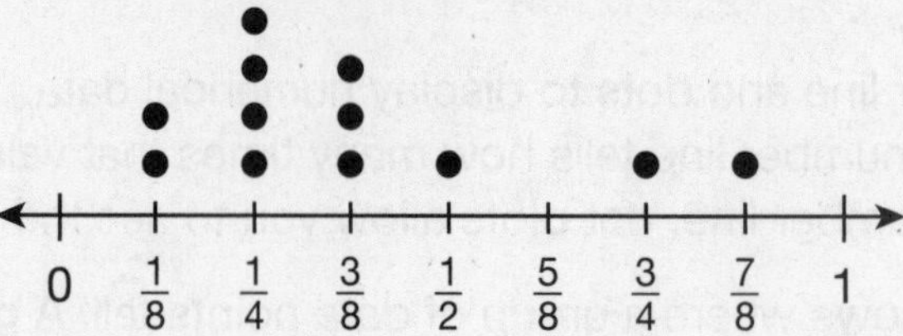

Step 6 Analyze the data.

There are a few values greater than or equal to $\frac{1}{2}$, but most of the values cluster from $\frac{1}{8}$ to $\frac{3}{8}$.

Solution **The dot plot for the weight data is shown in Step 3. The values cluster from $\frac{1}{8}$ pound to $\frac{3}{8}$ pound.**

Example 2

The dot plot below shows the ages of people in Tori's pottery class.

Ages

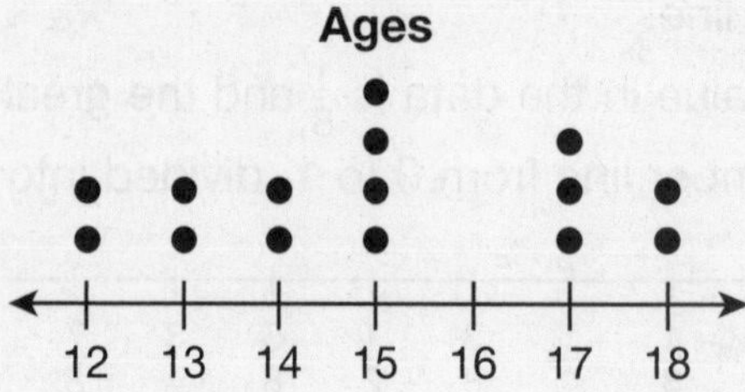

What are the mean, median, and mode ages of the people in the class?

Strategy **Analyze the dot plot.**

Step 1 Find the mean of the data.

First count the number of people in the class. There are 15 dots, so there are 15 people in the class.

Find the sum of the ages. Multiply each age by the number of dots above it, then add the products.

$(2 \times 12) + (2 \times 13) + (2 \times 14) + (4 \times 15) + (0 \times 16) + (3 \times 17) + (2 \times 18)$

$24 + 26 + 28 + 60 + 0 + 51 + 36 = 225$

Now divide that sum by the number of people in the class.

$225 \div 15 = 15$

The mean age is 15 years.

Step 2 Find the median of the data.

Divide the total number of dots by 2.

$15 \div 2 = 7.5$

Round up to the next whole number. 7.5 rounds to 8.

Find the value of the eighth dot.

Start at the first dot on the left and count to the eighth dot.

The eighth dot is above 15.

The median age is 15 years.

Step 3 Find the mode of the data.

The greatest number of dots is above 15.

The mode age is 15 years.

Solution **The mean, median, and mode ages are 15 years.**

Example 3

The dot plot shows the distance that the students in Miss Hall's class live from school.

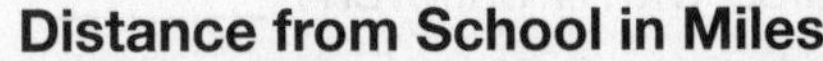

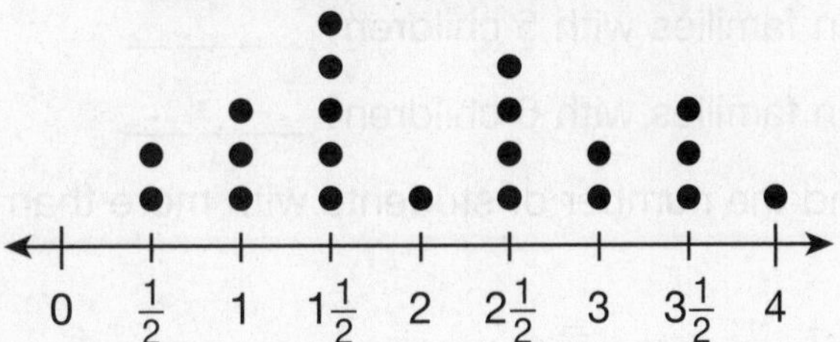

What is the range of the distances that the students live from school?

Strategy **Use the dot plot.**

Step 1 Find the greatest data value.

The greatest number of dots is above 4.

Step 2 Find the least data value.

The least number of dots is above $\frac{1}{2}$.

Step 3 Find the difference between the greatest and least values.

$4 - \frac{1}{2} = 3\frac{1}{2}$

Solution **The range is $3\frac{1}{2}$ miles.**

Coached Example

Jeffrey surveyed all the students in his class. He asked them how many children were in their families. He recorded the results in this dot plot.

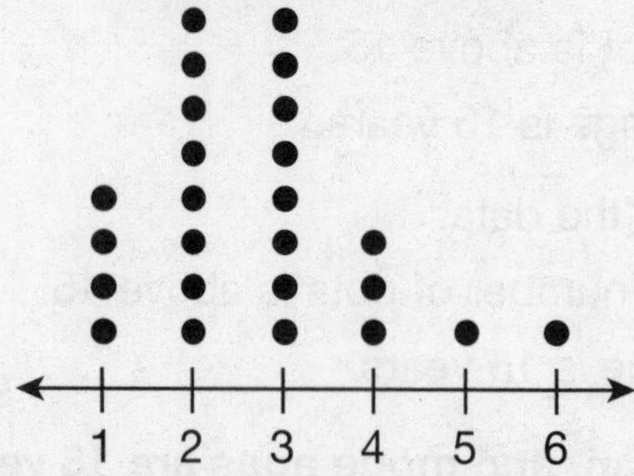

How many students are in families with more than 3 children?

More than 3 children means _______ or more children.

How many students are in families with 4 children? _______

How many students are in families with 5 children? _______

How many students are in families with 6 children? _______

Add those numbers to find the number of students with more than 3 children in their families.

_______ + _______ + _______ = _______

A total of _______ students have families with more than 3 children.

Lesson Practice

Choose the best answer.

Use the dot plot for questions 1–3.

Javier recorded the high temperature each day in the dot plot below.

Daily High Temperatures (in °F)

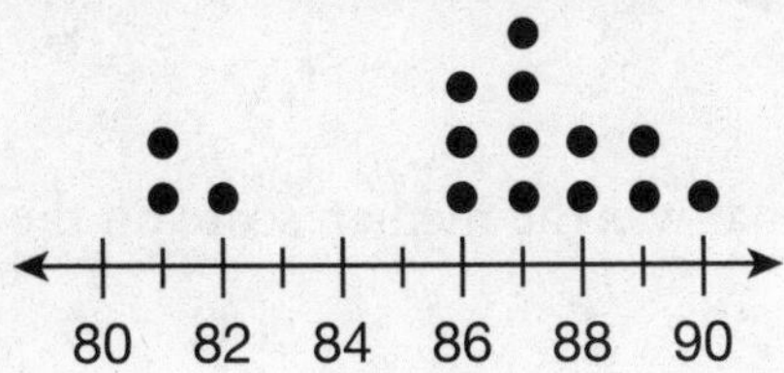

1. On how many days did Javier record the high temperature?

   A. 1
   B. 9
   C. 15
   D. 90

2. What was the mode temperature during the sample period?

   A. 90°F
   B. 89°F
   C. 88°F
   D. 87°F

3. Between which temperatures is there a gap in the data?

   A. between 80°F and 82°F
   B. between 82°F and 86°F
   C. between 85°F and 91°F
   D. between 89°F and 91°F

Use the dot plot for questions 4–6.

The dot plot below shows how Vicki split up a large bag of rice into smaller bags.

Bags of Rice in Pounds

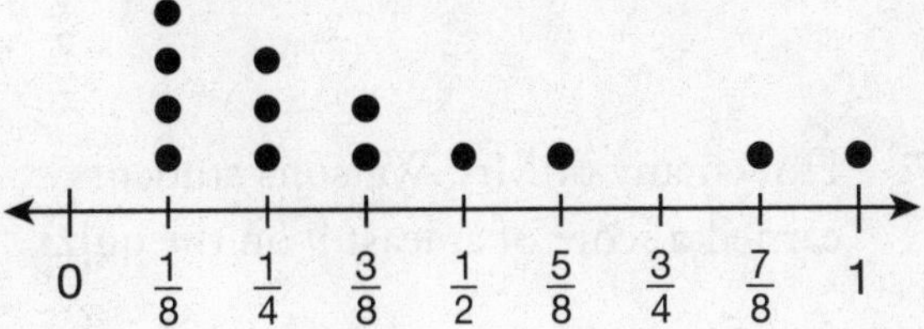

4. What is the mode of the data?

   A. $\frac{1}{8}$ pound
   B. $\frac{1}{4}$ pound
   C. $\frac{3}{8}$ pound
   D. no mode

5. What is the median of the data?

   A. $\frac{1}{8}$ pound
   B. $\frac{1}{4}$ pound
   C. $\frac{3}{8}$ pound
   D. $\frac{1}{2}$ pound

6. Which statement is **not** true?

   A. The data clusters from $\frac{1}{8}$ pound to $\frac{3}{8}$ pound.
   B. Fewer than half the bags weigh more than $\frac{1}{2}$ pound.
   C. The range of the weights is $\frac{7}{8}$ pound.
   D. There is no gap in the data.

Use the dot plot for questions 7 and 8.

The dot plot below shows the scores from Mrs. Watson's most recent spelling quiz.

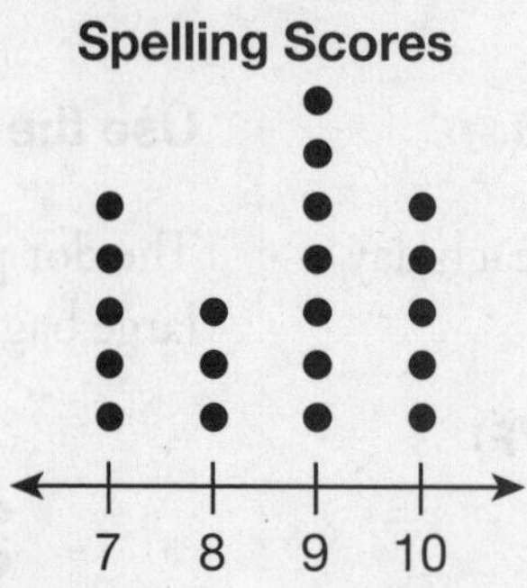

7. How many of Mrs. Watson's students earned a score of at least 9 on the quiz?

   A. 5
   B. 7
   C. 12
   D. 15

8. What was the median score on the quiz?

   A. 7
   B. 8
   C. 9
   D. 10

9. The following data set gives the hourly wages of the workers at a manufacturing plant.
   \$11, \$10, \$8, \$12, \$10, \$13, \$9, \$10, \$12, \$15, \$10, \$12, \$10, \$12, \$11

   A. Make a dot plot of the data.

   Hourly Wages (in dollars)

   B. Are there any clusters of data? Explain your answer.

____

____

____

Common Core State Standards: 6.SP.5.a, 6.SP.5.d

Choose the Best Measure

Getting the Idea

The shape of the data in a data set can help you choose whether to use the median or the mean as a measure of center.

Example 1

Mr. Spector separated his class into groups to do a science project. The dot plot below shows the number of hours that each group worked on its science project.

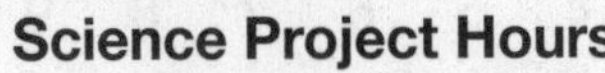

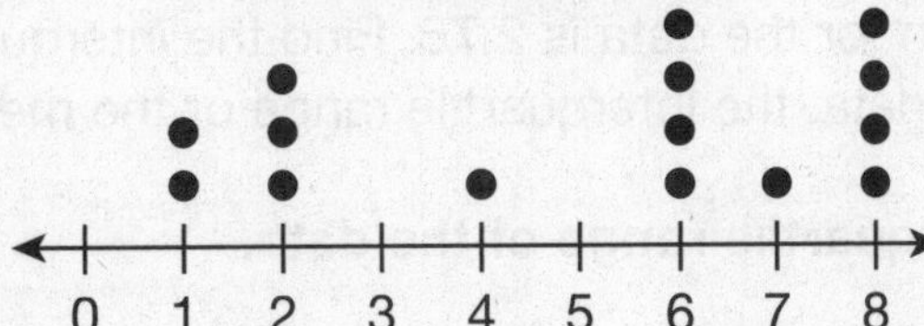

Which measure of center, the median or the mean, is the better measure for describing the data?

Strategy **Find the median and the mean of the data in the line plot.**

Step 1 Find the median.

Count the number of dots on the line plot.

There are 15 dots, so there were 15 groups in the class.

$15 \div 2 = 7.5$

The median will be the value of the eighth dot.

Starting at the left, count dots until you get to the eighth dot.

The median is 6.

Step 2 Find the mean.

Add: $1 + 1 + 2 + 2 + 2 + 4 + 6 + 6 + 6 + 6 + 7 + 8 + 8 + 8 + 8 = 75$

Divide by 15: $75 \div 15 = 5$

The mean is 5.

Step 3 Look at the shape of the data.

Most of the data is greater than or equal to 6, the median.

Solution **The median is a better measure for describing the data.**

The shape of the data in a data set can also affect whether you use the interquartile range or the mean absolute deviation as a measure of variability.

Example 2

The booster club is selling sweatshirts for a fundraiser. The dot plot shows how many sweatshirts were sold by each member of the booster club on the first day.

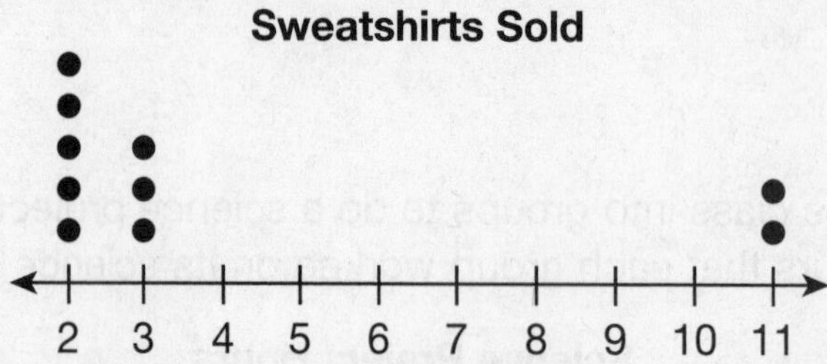

The mean absolute deviation for the data is 2.76. Find the interquartile range for the data. Which better describes the data, the interquartile range or the mean absolute deviation?

Strategy **Find the interquartile range of the data.**

Step 1 Find the median of the data.

Order the data values from least to greatest.

2, 2, 2, 2, 2, 3, 3, 3, 11, 11

The two middle values are 2 and 3, so the median is 2.5.

Step 2 Find the first and third quartiles.

The first quartile is the median of the lower half of the data: 2, 2, 2, 2, 2.

The middle number is 2, so the first quartile is 2.

The third quartile is the median of the upper half of the data: 3, 3, 3, 11, 11.

The middle number is 3, so the third quartile is 3.

Step 3 Find the interquartile range.

The interquartile range is the difference of the first and third quartiles.

$3 - 2 = 1$

The interquartile range of the data is 1.

Step 4 Look at the shape of the data.

Most of the data is clustered around 2 and 3. The difference of 3 and 2 is 1, which is the same as the interquartile range.

The data does not show as much variability as the mean absolute deviation.

Solution **The interquartile range of the data is 1. The interquartile range describes the data better than the mean absolute deviation does.**

Coached Example

The dot plot shows the number of pairs of sneakers of each size sold at a shoe store in one day.

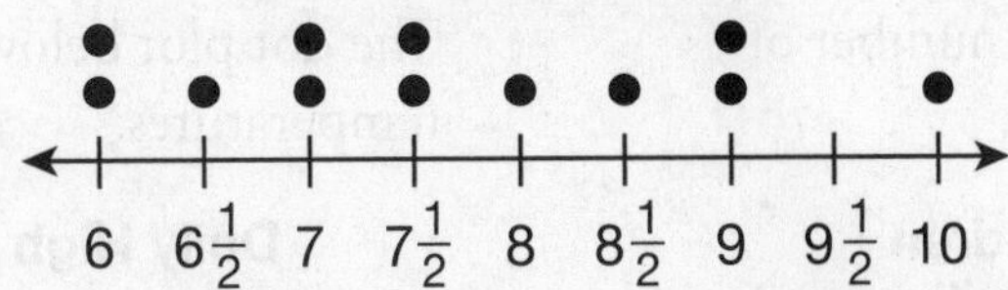

Which measure of center, the median or mean, better describes the data? Explain your answer.

Find the median.

Count the number of dots on the line plot. There are ______ dots.

12 ÷ 2 = ______

The median will be the mean of the sixth and seventh dots.

The value of both the sixth and seventh dots is ______. The median is ______.

Find the mean. Add the values.

______ + ______ + ______ + ______ + ______ + ______ + ______ +
______ + ______ + ______ + ______ + ______ = ______

Divide by ______. Write the answer as a mixed number.

______ ÷ ______ = ______

The mean is ______.

Do shoe sizes come in the same size as the mean?

The ______ is a better measure for describing the data.

Lesson Practice

Choose the best answer.

Use the dot plot for questions 1–3.

The dot plot below show the number of children in students' families.

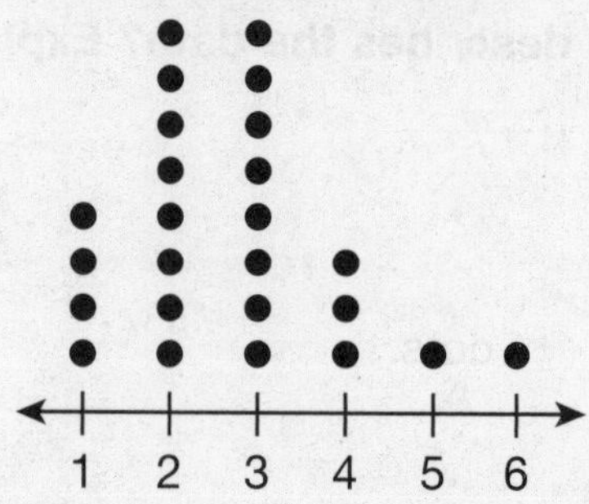

1. What is the mean of the data?

   A. 2.10
   B. 2.68
   C. 2.75
   D. 3.25

2. What is the median of the data?

   A. 2
   B. 3
   C. 4
   D. 5

3. Which statement best describes the data?

   A. The mean best describes the data because more students have at least 4 children in their families.
   B. The mean best describes the data because 3 is also the mode.
   C. The median best describes the data because the data is clustered around 2 and 3.
   D. The median best describes the data because more students have fewer than 3 children in their families.

Use the dot plot for questions 4 and 5.

The dot plot below show daily high temperatures.

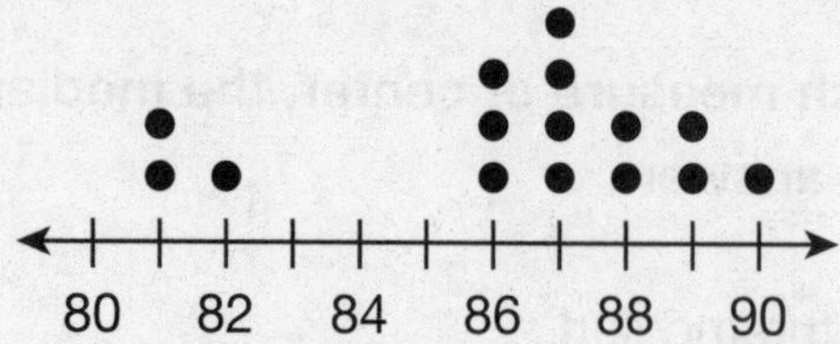

4. What is the interquartile range (IQR) of the data?

   A. 9
   B. 4
   C. 2
   D. 1

5. The mean absolute deviation (MAD) of the data is 2.08. Which statement best describes the data?

   A. The MAD describes the data better than the IQR because the data has no clusters.
   B. The MAD describes the data better than the IQR because it has a lesser value.
   C. The IQR describes the data better than the MAD because it has a greater value.
   D. The IQR describes the data better than the MAD because the data is clustered.

6. The following low temperatures were recorded for 10 days in February.

15°F, 18°F, 10°F, 17°F, 17°F, 20°F, 16°F, 17°F, 15°F, 14°F

A. Construct a dot plot for the temperatures.

Low Temperatures (in °F)

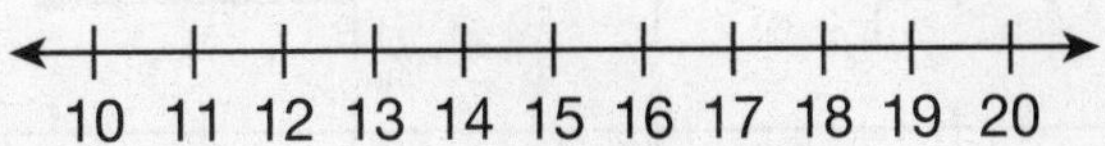

B. Which measure of center best describes the data, the median or the mean? Explain your answer.

Box Plots

Getting the Idea

A **box plot** can help you see the variation in a data set. You can see some measures of variation easily on a box plot, as shown below.

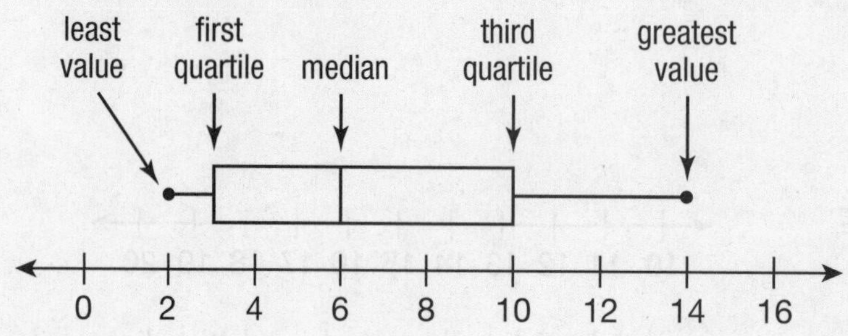

Example 1

Make a box plot of the following data.

250, 175, 215, 350, 320, 235, 250, 280

Strategy **Find the least and greatest values, the quartiles, and the median.**

Step 1 Order the numbers from least to greatest.

175, 215, 235, 250, 250, 280, 320, 350

The least value is 175. The greatest value is 350.

Step 2 Find the median.

175, 215, 235, <u>250</u>, <u>250</u>, 280, 320, 350

The median is 250.

Step 3 Find the first quartile.

The first quartile is the median of the lower half of the data.

175, <u>215</u>, <u>235</u>, 250

$215 + 235 = 450$ and $450 \div 2 = 225$.

The first quartile is 225.

Step 4 Find the third quartile.

The third quartile is the median of the upper half of the data.

250, <u>280</u>, <u>320</u>, 350

$280 + 320 = 600$ and $600 \div 2 = 300$.

The upper quartile is 300.

Step 5 Make the box plot.

Pick an interval that makes the box plot easy to read.

All of the values are multiples of 25, so use intervals of 25.

Use dots for the least and greatest values.

Use line segments for the quartiles and the median.

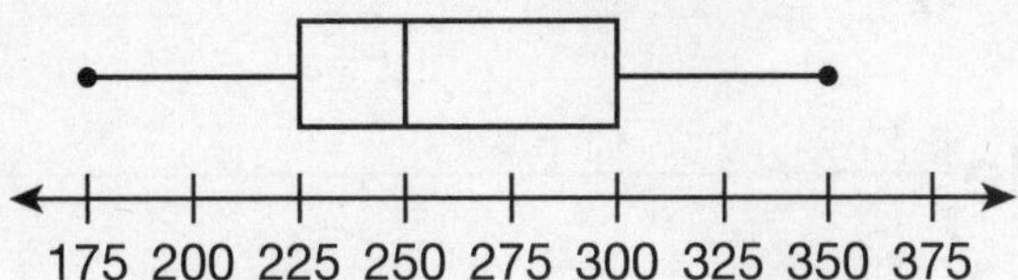

Solution **A box plot of the data is shown in Step 5 above.**

Example 2

Isabella drew the box plot below to show the heights, in inches, of her tomato plants.

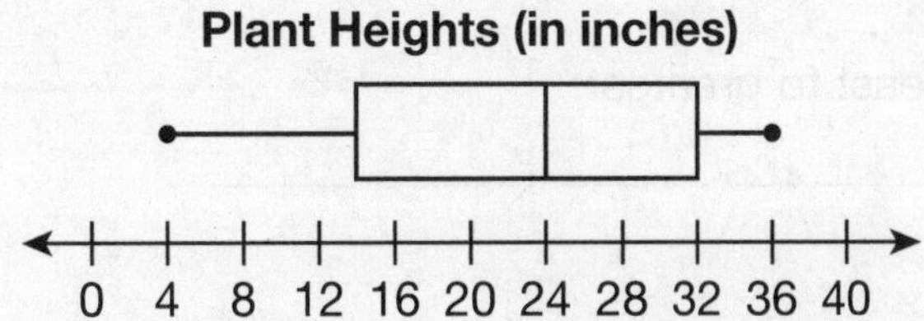

What are the median plant height and the interquartile range, in both inches and feet?

Strategy **Find the median and interquartile range in inches. Then convert to feet.**

Step 1 Identify the median.

The line segment inside the box represents the median.

The median is 24 inches.

Step 2 Convert the height of the median to feet.

12 inches = 1 foot, so 24 inches = 2 feet.

The median plant height is 2 feet.

Step 3 Find the interquartile range.

The interquartile range is the difference of the third quartile and the first quartile.

The third quartile is 32. The first quartile is 14.

$32 - 14 = 18$

The interquartile range is 18 inches.

Step 4 Convert the interquartile range to feet.

12 inches = 1 foot, so 18 inches = 1 foot 6 inches or $1\frac{1}{2}$ feet.

The interquartile range is $1\frac{1}{2}$ feet.

Solution **The median plant height is 24 inches or 2 feet. The interquartile range is 18 inches or $1\frac{1}{2}$ feet.**

Coached Example

Molly recorded the heights, in inches, of her cousins. Make a box plot of the data.

47, 56, 34, 52, 44, 28, 38

To make a box plot, you need to find five pieces of data: ________________ value, ____________ quartile, median, ____________ quartile, and ____________ value.

Order the numbers from least to greatest:

_____, _____, _____, _____, _____, _____, _____

Least value: ________

Greatest value: ________

The median is the ______________ number in the ordered list.

Median: ________

The first quartile is the median of the numbers in the ____________ half of the data.

First quartile: ________

The third quartile is the median of the numbers in the ____________ half of the data.

Third quartile: ________

Complete the number line below. Then make your box plot.

Heights (in inches)

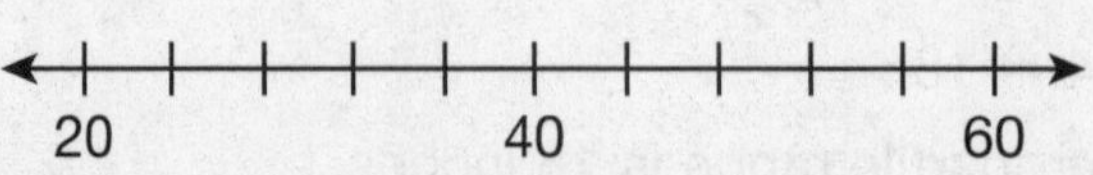

A box plot for the data is shown above.

Lesson Practice

Choose the correct answer.

Use the data for questions 1–4.

The data shows the ages of teachers at a school. You may draw a box plot to help you.

33, 25, 42, 25, 31, 37, 43, 29, 38, 25, 25, 29, 31, 33, 37, 38, 42, 43

1. What is the median of the data?

   A. 25
   B. 31
   C. 33
   D. 37

2. What is the first quartile of the data?

   A. 29
   B. 30
   C. 31
   D. 33

3. What is the third quartile of the data?

   A. 33
   B. 37
   C. 38
   D. 40

4. What is the interquartile range of the data?

   A. 8
   B. 9
   C. 10
   D. 18

Use the box plot for questions 5–8.

The box plot shows the heights, in inches, of the people in Joe's family.

Heights (in inches)

56 58 60 62 64 66 68 70 72

5. What does 66 represent in the box plot?

   A. least value
   B. median
   C. mean
   D. greatest value

6. What does 69 represent in the box plot?

   A. mean
   B. median
   C. first quartile
   D. third quartile

7. What is the interquartile range of the data, in feet?

   A. $\frac{1}{4}$ foot
   B. $\frac{1}{2}$ foot
   C. $\frac{2}{3}$ foot
   D. $1\frac{1}{3}$ feet

8. Suppose the data set that created this box plot was as follows:

   56, 61, 61, 66, 69, 69, 72

   Which of the following would happen to the box plot if 70 were added to the data?

   A. The first quartile would change.
   B. The median would change.
   C. The greatest value would increase.
   D. The least value would decrease.

9. The data below represents the bowling scores of a group of friends.

93, 129, 102, 115, 136, 85, 134, 108, 112, 125, 150, 144

A. Find the least value, first quartile, median, third quartile, and greatest value for the data.

least value: ___________

greatest value: ___________

first quartile: ___________

median: ___________

third quartile: ___________

B. Make a box plot of the data.

Common Core State Standards: 6.SP.4, 6.SP.5.a, 6.SP.5.c

Histograms

Getting the Idea

Frequency is a measure of how many times an event occurs. A **frequency table** can be used to record and display data. A frequency table may use **tally marks**. Each | represents 1 and each 𝍸 represents 5.

You can display data from a frequency table in a histogram. A **histogram** is a bar graph that shows the frequency of data within equal **intervals**. A histogram does not have gaps between the bars, unless the frequency for an interval is 0.

Example 1

Rex and Jai performed a science experiment in which they recorded the outdoor temperature for 30 consecutive mornings. They displayed the results in the frequency table below.

Morning Temperature

Temperature (in °F)	Tally	Frequency
15–19	𝍸 𝍸 ||	12
20–24	𝍸 𝍸 ||||	14
25–29	|||	3
30–34	|	1

Display the data in a histogram.

Strategy **Use the data in the frequency table to make the histogram.**

Step 1 Decide how to label the axes.

Use the horizontal axis for the temperature intervals and the vertical axis for the frequency.

Step 2 Draw bars indicating the frequency for each of the intervals.

Step 3 Give the histogram a title.

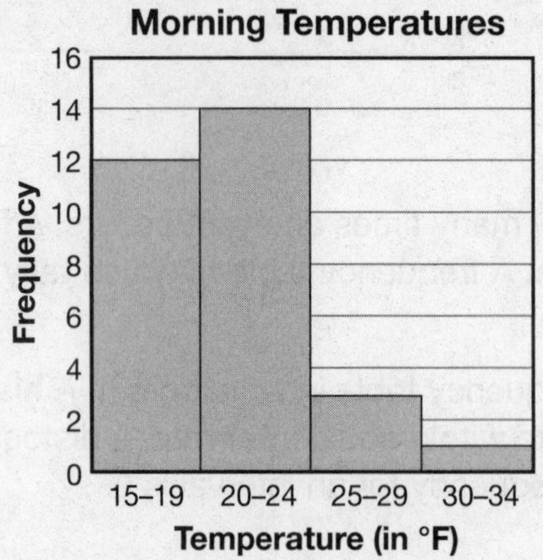

Solution **A histogram for the temperature data is shown in Step 3 above.**

Example 2

The histogram shows the number of phone calls that Sherry made each day in April.

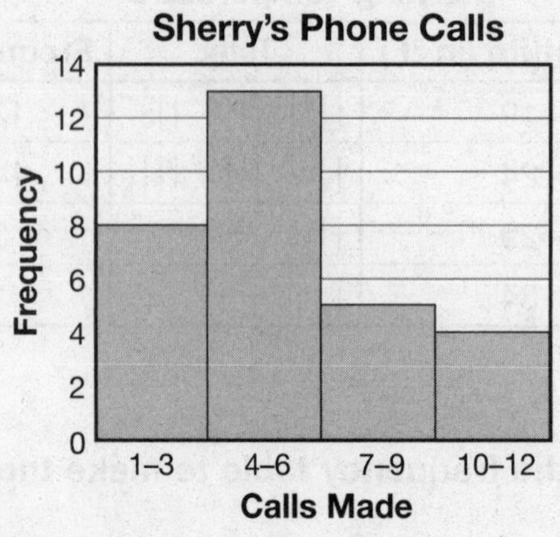

On how many days did Sherry make from 1 to 6 calls?

Strategy **Find the frequencies for the intervals 1–3 and 4–6.**

Step 1 Find the frequency for 1–3 calls.

The bar for 1–3 calls stops at 8.

Sherry made 1–3 calls on 8 different days.

Step 2 Find the frequency for 4–6 calls.

The bar for 4–6 calls stops halfway between 12 and 14, or at 13.

Sherry made 4–6 calls on 13 different days.

Step 3 Add the frequencies for the two intervals.

$8 + 13 = 21$

Solution **Sherry made from 1 to 6 calls on 21 days in April.**

Example 3

The histogram below shows the times, in seconds, of the students on the track team in the 200-meter dash.

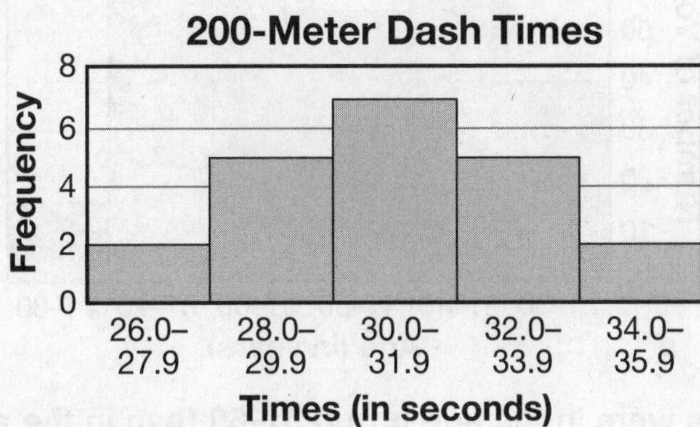

What time interval contains the median time?

Strategy **Find the total number of data items. Then find the interval that contains the median.**

Step 1 Find the number of students who ran the 200-meter dash.

Each bar in the histogram tells you how many students had a 200-meter dash time that fell in that interval.

Add the frequencies for the intervals.

$2 + 5 + 7 + 5 + 2 = 21$

Twenty-one students ran the 200-meter dash.

Step 2 Find the interval that contains the median.

There are 21 data items.

$21 \div 2 = 10.5$, so the median is the 11th data item in an ordered list.

The data in the histogram is also ordered from least to greatest.

Starting at the left, add the frequencies until you get to the interval that contains the 11th data item.

$2 + 5 = 7$ and $2 + 5 + 7 = 14$

The median is in the third interval, the times from 30 to 31.9 seconds.

Solution **The median time is in the interval for 30 to 31.9 seconds.**

Coached Example

The histogram below shows the number of voters by age group in an election for mayor.

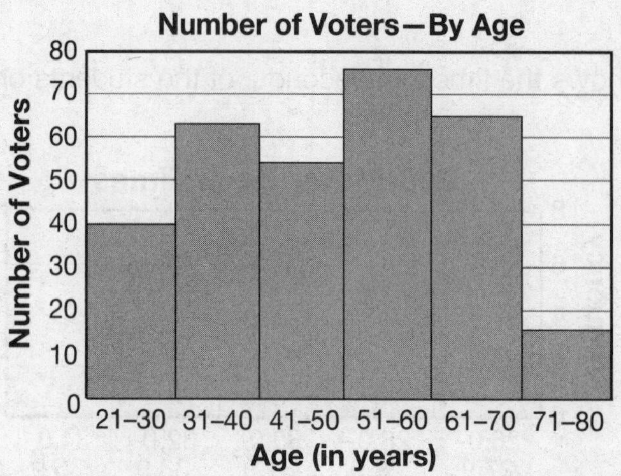

How many more voters were in the age group 51–60 than in the age group 21–30?

Find the number of voters in the age group 51–60.

The bar for the age group 51–60 is halfway between ____________ and ____________.

There are ____________ voters in the age group 51–60.

Find the number of voters in the age group 21–30.

The bar for the age group 21–30 ends on ____________.

Subtract. ____________ – ____________ = ____________

There were ____________ more voters in the age group 51–60 than in the age group 21–30.

Lesson Practice

Choose the correct answer.

Use the histogram below for questions 1–3.

1. How many people spent from $30.00 to $39.99?

   A. 80   C. 45
   B. 60   D. 15

2. How many more people spent less than $20 than spent at least $20?

   A. 20
   B. 60
   C. 80
   D. 100

3. How many people in all spent money at the raffle?

   A. 200   C. 80
   B. 140   D. 60

Use the histogram below for questions 4 and 5.

Ralph surveyed his neighbors about the number of bikes they own. The results are in the histogram below.

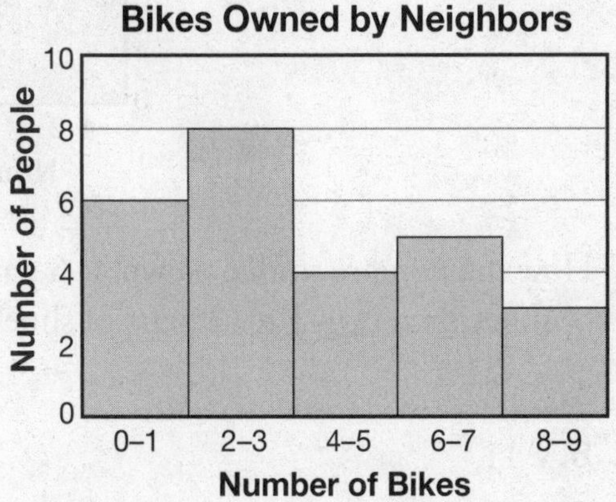

4. How many people did Ralph survey?

   A. 10
   B. 14
   C. 26
   D. 30

5. How many more people own fewer than 4 bikes than own more than 5 bikes?

   A. 2
   B. 6
   C. 8
   D. 14

Use the histogram below for questions 6 and 7.

The histogram shows the number of pairs of shoes owned by sophomores at Ferris High School.

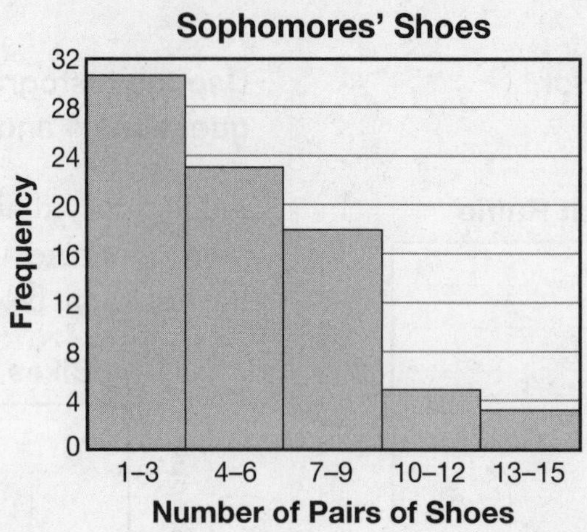

6. How many more students own 4–6 pairs of shoes than own 10–12 pairs of shoes?

   A. 5

   B. 11

   C. 18

   D. 23

7. How many fewer students own 13–15 pairs of shoes than own 1–3 pairs of shoes?

   A. 3

   B. 13

   C. 24

   D. 28

8. Students' scores on Ms. Rivera's last math test are shown below.

82, 96, 91, 100, 94, 78, 100, 90, 95, 88, 92, 98,
100, 82, 93, 80, 94, 90, 76, 90, 84, 100, 82, 96

A. Use the data above to complete the frequency table below.

Ms. Rivera's Math Test Scores

Score	Tally	Frequency
76–80		
81–85		
86–90		
91–95		
96–100		

B. Make a histogram of the data.

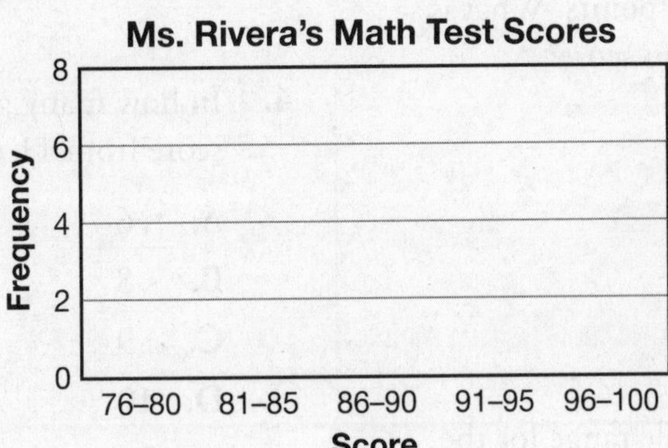

Domain 5: Cumulative Assessment for Lessons 32–37

1. Which of the following questions is a statistical question?

   A. How much money is my pocket?

   B. How many students take the bus to school?

   C. What is my house number?

   D. How many pets does each student in your class have?

2. In its first 5 games, a football team scored 14, 10, 17, 13, and 21 points. What is the football team's mean score?

   A. 11

   B. 14

   C. 15

   D. 17

3. What is the interquartile range for the data on the box plot?

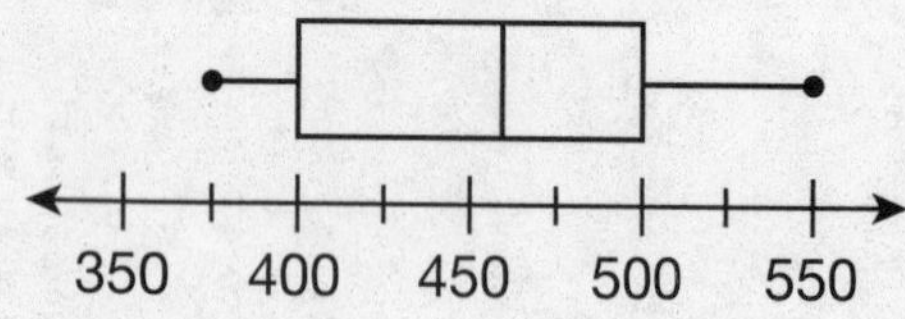

   A. 100

   B. 150

   C. 175

   D. 200

Use the histogram for questions 4 and 5.

The histogram shows the number of points that the Panthers football team scored in each game this season.

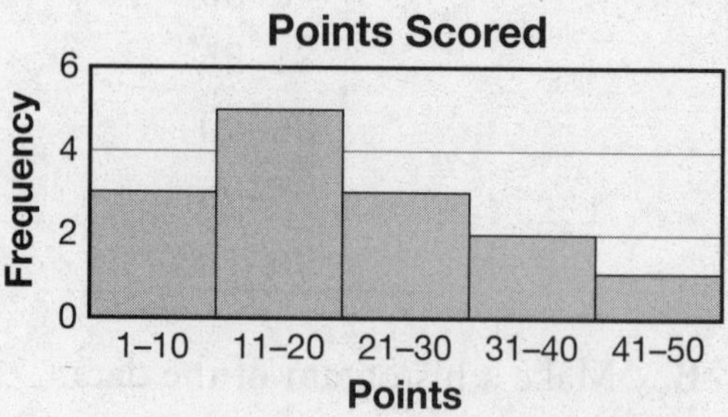

4. In how many games did the Panthers score from 11 to 30 points?

   A. 6

   B. 8

   C. 9

   D. 10

5. The Panthers won every time they scored at least 21 points. They won half of their other games. How many games did the Panthers win?

   A. 8

   B. 10

   C. 12

   D. 14

Use the dot plot for questions 6 and 7.

The dot plot shows the number of books that the students in Ms. Wilson's class read independently during the school year.

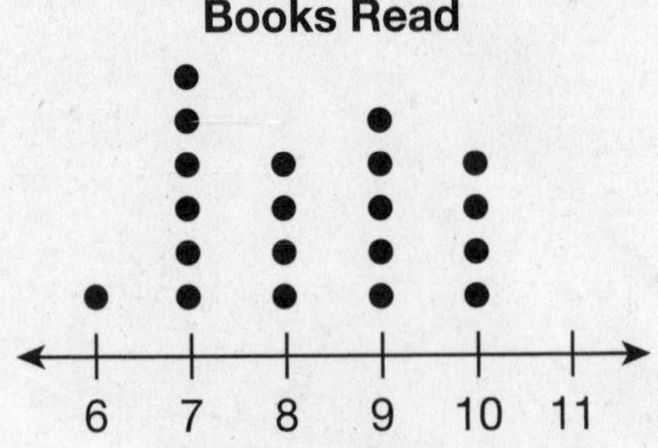

6. How many students read more than 8 books independently?

A. 9

B. 11

C. 13

D. 19

7. Which statement is **not** true?

A. The data clusters from 7 to 10 books.

B. The mode is 7 books.

C. There are 20 students in Ms. Wilson's class.

D. There is a gap in the data.

8. For which set of data would the mean **not** be best to describe the data?

A. the incomes of families in a city

B. the selling price of a gallon of milk

C. the sizes of sweaters sold in a store

D. the miles per gallon of all the models made by a car company

9. The following data set shows the number of seashells collected on the beach by a group of friends.

25, 17, 20, 19, 21, 24,
22, 18, 21, 19, 20, 23

Make a dot plot of the data.

Seashells Collected

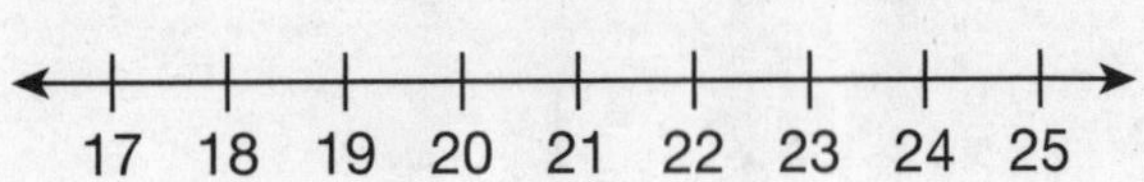

10. The high temperatures for five days last week were 64°F, 71°F, 70°F, 68°F, and 72°F.

A. Find the interquartile range of the data set. Show your work.

____

B. How does the interquartile range reflect the temperature data?

____

____

____

Glossary

absolute value the distance of a number from zero on a number line (Lesson 4)

area a measure of the number of square units needed to cover a region (Lesson 24)

base a face of a solid figure; the face for which a solid figure is named (Lesson 29)

box plot a graph that shows the least and greatest values, first and third quartiles, and median of a data set (Lesson 36)

capacity the amount that a container can hold (Lesson 16)

centimeter (cm) a metric unit for measuring length; 100 centimeters = 1 meter (Lesson 16)

cluster a group of items in a data set that are close in value (Lesson 34)

coefficient the numerical factor of a term containing a variable (Lesson 17)

composite polygon a polygon that can be divided into simpler figures (Lesson 26)

coordinate plane a plane formed by the intersection of a horizontal number line, the x-axis, and a vertical number line, the y-axis (Lesson 10)

cross multiply to determine whether two ratios are equivalent by multiplying the numerator of each ratio by the denominator of the other ratio (Lesson 13)

cross product the product of the numerator of one ratio and the denominator of another ratio (Lesson 13)

cubic unit a cube with an edge length of 1 of any given unit (Lesson 31)

cup (c) a customary unit of capacity; 1 cup = 8 fluid ounces (Lesson 16)

customary system the system of measurement used in the United States (Lesson 16)

dependent variable in an equation of two variables, the variable whose value is affected by the independent variable; in $y = x + 1$, y is the dependent variable (Lesson 21)

distributive property a property that states that multiplying the sum of two numbers by another number gives the same result as multiplying each addend by the number and then adding the products; the property also applies to subtraction (Lesson 1)

dividend a number to be divided (Lesson 2)

divisor the number by which the dividend is divided (Lesson 2)

dot plot a data display in which each data item is shown as a dot above a number line (Lesson 34)

edge a line segment where two faces of a solid figure meet (Lesson 29)

equation a statement that two quantities are equal; contains an equal sign (=) (Lesson 20)

equivalent expressions expressions that name the same number for any value of a variable (Lesson 19)

equivalent ratios ratios $\frac{a}{b}$ and $\frac{c}{d}$ are equivalent if $ad = bc$, where $b \neq 0$ and $d \neq 0$ (Lesson 13)

expression a mathematical statement that combines numbers, operation signs, and sometimes variables (Lesson 17)

face a flat surface of a solid figure (Lesson 29)

factor (of a number) a number that can be multiplied by another number to get a given number, or a number that divides evenly into a given number (Lesson 1)

first quartile the median of the lower half of a data set (Lesson 33)

fluid ounce (fl oz) a customary unit of capacity; 8 fluid ounces = 1 cup (Lesson 16)

foot (ft) a customary unit for measuring length; 1 foot = 12 inches (Lesson 16)

frequency a measure of how many times an event occurs (Lesson 37)

frequency table a data display that shows how often each item occurs in a data set (Lesson 37)

gallon (gal) a customary unit of capacity; 1 gallon = 4 quarts (Lesson 16)

gap in a dot plot, an interval where there are no data items (Lesson 34)

gram (g) a metric unit of mass; 1 gram = 1,000 milligrams (Lesson 16)

greatest common factor (GCF) the greatest factor that is common to two or more numbers (Lesson 1)

histogram a bar graph that shows the frequency of data within equal intervals (Lesson 37)

inch (in.) a customary unit for measuring length; 12 inches = 1 foot (Lesson 16)

independent variable in an equation of two variables, the variable whose value affects the dependent variable; in $y = x + 1$, x is the independent variable (Lesson 21)

inequality a mathematical statement that compares two expressions and includes an inequality symbol such as $<$, $>$, $\leq$, or $\geq$ (Lesson 23)

integers the set of counting numbers (1, 2, 3, . . .), their opposites (−1, −2, −3, . . .), and zero (Lesson 3)

interquartile range (IQR) the difference of the first and third quartiles in a data set (Lesson 33)

interval a range of values (Lesson 37)

inverse operations operations that undo each other, such as addition and subtraction or multiplication and division (Lessons 9, 20)

is equal to (=) a symbol that shows that two quantities have the same value (Lesson 6)

is greater than (>) a symbol that compares two quantities; it shows that the first quantity has a greater value than the second quantity (Lessons 6, 23)

is greater than or equal to (≥) a symbol that compares two quantities; it shows that the first quantity has a value greater than or equal to the value of the second quantity (Lesson 23)

is less than (<) a symbol that compares two quantities; it shows that the first quantity has a lesser value than the second quantity (Lessons 6, 23)

is less than or equal to (≤) a symbol that compares two quantities; it shows that the first quantity has a value less than or equal to the value of the second quantity (Lesson 23)

kilogram (kg) a metric unit of mass; 1 kilogram = 1,000 grams (Lesson 16)

kilometer (km) a metric unit for measuring length; 1 kilometer = 1,000 meters (Lesson 16)

lateral face a face that is not the base in a pyramid or a prism (Lesson 30)

least common multiple (LCM) the least number that is a multiple of two or more numbers (Lesson 1)

length a measure of how long or tall something is (Lesson 16)

like terms terms that have the same variable raised to the same power (Lesson 19)

liter (L) a metric unit of capacity; 1 liter = 1,000 milliliters (Lesson 16)

mass the amount of matter in an object (Lesson 16)

mean the sum of the values in a data set divided by the number of values in the data set (Lesson 32)

mean absolute deviation (MAD) a measure of the amount of variability in a data set; the mean of the absolute values of the deviations from the mean in a data set (Lesson 33)

measure of center a single number that describes all the values in a data set; a number that describes the middle or average of a data set (Lesson 32)

measure of variability a single number that describes all the values in a data set; a single number that describes how spread out or how clustered a set of data is (Lesson 33)

median the middle value in a data set ordered from least to greatest (Lesson 32)

meter (m) a metric unit for measuring length; 1 meter = 100 centimeters (Lesson 16)

metric system a measurement system used in most of the world other than the United States; it is based on powers of 10 (Lesson 16)

metric ton (t) a metric unit of mass; 1 metric ton = 1,000 kilograms (Lesson 16)

mile (mi) a customary unit for measuring length; 1 mile = 1,760 yards or 5,280 feet (Lesson 16)

milligram (mg) a metric unit of mass; 1,000 milligrams = 1 gram (Lesson 16)

milliliter (mL) a metric unit of capacity; 1,000 milliliters = 1 liter (Lesson 16)

millimeter (mm) a metric unit of length; 10 millimeters = 1 centimeter (Lesson 16)

mode the value or values that occur most often in a data set (Lesson 32)

multiple (of a number) the product of a given number and any counting number (Lesson 1)

negative integers integers less than zero (Lesson 3)

net a flat pattern that can be folded into a three-dimensional figure; it shows each surface of the solid figure it forms (Lesson 29)

opposites numbers that are the same distance from 0 on a number line (Lesson 3)

order of operations a set of rules that determines the correct sequence for evaluating expressions (Lesson 18)

ordered pair a pair of numbers in the form (*x*, *y*) that gives the location of a point (Lesson 10)

origin the point on a coordinate plane where the *x*-axis and the *y*-axis intersect, located at (0, 0) (Lesson 10)

ounce (oz) a customary unit of weight; 16 ounces = 1 pound (Lesson 16)

outlier a value that is much greater than or much less than the other values in a data set (Lesson 32)

percent a ratio that compares a number to 100; "per 100" or "out of 100" (Lesson 15)

perimeter the distance around the outside of a closed figure (Lesson 11)

pint (pt) a customary unit of capacity; 1 pint = 2 cups (Lesson 16)

positive integers integers greater than zero (Lesson 3)

pound (lb) a customary unit of weight; 1 pound = 16 ounces (Lesson 16)

prism a solid figure with two parallel bases that are congruent polygons whose other faces are rectangles or parallelograms (Lesson 29)

pyramid a solid figure whose base is a polygon and whose other faces are triangles (Lesson 29)

quadrant any of the four sections of a coordinate plane separated by the *x*-axis and *y*-axis (Lesson 10)

quadrilateral a polygon with 4 sides (Lesson 25)

quart (qt) a customary unit of capacity; 1 quart = 2 pints (Lesson 16)

quotient the answer to a division problem (Lesson 2)

range the difference of the least value and the greatest value in a data set (Lesson 33)

rate a ratio that compares two quantities with different units of measure (Lesson 14)

ratio a comparison of two numbers using division (Lesson 12)

rational number a number that can be expressed as the ratio of two integers in the form $\frac{a}{b}$, where *b* is not equal to 0 (Lesson 5)

reciprocals two numbers whose product is 1; *x* and $\frac{1}{x}$ are reciprocals (Lesson 9)

reflection a flip of a point or figure, usually over a line such as an axis on a coordinate plane (Lesson 11)

regular polygon a polygon with all sides and all angles congruent (Lesson 25)

remainder an amount that is left over when a number is divided by another number; always less than the divisor (Lesson 2)

slant height the height of a lateral face in a pyramid (Lesson 30)

solid figure any figure that has length, width, and height; also called a three-dimensional figure (Lesson 29)

solution set the numbers that are solutions to an inequality (Lesson 23)

square unit a square with a side length of 1 of any given unit (Lesson 24)

surface area the total area of the surfaces of a solid figure (Lesson 30)

tally mark a mark used to record the frequency of data (Lesson 37)

term a number, variable, product, or quotient in an expression (Lesson 17)

third quartile the median of the upper half of a data set (Lesson 33)

three-dimensional figure any figure that has length, width, and height; also called a solid figure (Lesson 29)

ton (T) a customary unit of weight; 1 ton = 2,000 pounds (Lesson 16)

unit rate a rate in which the second measurement or amount is 1 unit (Lesson 14)

variable a letter or symbol used to represent a number (Lesson 17)

vertex the point where 3 or more edges meet in a solid figure (Lesson 29)

volume a measure of how much space a solid figure takes up; a measure of the number of cubic units that fit inside a solid figure (Lesson 31)

weight the gravitational force on an object (Lesson 16)

***x*-axis** the horizontal axis on a coordinate plane (Lesson 10)

***x*-coordinate** the first number of an ordered pair of numbers (Lesson 10)

***y*-axis** the vertical axis on a coordinate plane (Lesson 10)

***y*-coordinate** the second number of an ordered pair of numbers (Lesson 10)

yard (yd) a customary unit for measuring length; 1 yard = 3 feet or 36 inches (Lesson 16)

Crosswalk Coach for the Common Core State Standards
Mathematics, Grade 6

Summative Assessment: Domains 1–5

Name: ______________________________

Session 1

1. Three seedless watermelons cost $27. If Pablo buys 4 seedless watermelons at that same unit price, how much will he pay?

   A. $9

   B. $32

   C. $36

   D. $108

2. What is the least common multiple of 5 and 6?

   A. 15

   B. 18

   C. 25

   D. 30

3. Muna has a piece of string that is $4\frac{2}{3}$ yards long. If she cuts it into pieces that are each $\frac{2}{3}$ yard long, how many pieces of string will she have?

   A. 14

   B. 7

   C. 4

   D. 3

4. A soccer coach wants to order soccer balls online for her team. The soccer balls cost $9 each and there is a shipping charge of $10. If s represents the number of soccer balls that the coach buys, which expression can be used to find the total cost of the soccer balls?

   A. $\frac{s}{9} + 10$

   B. $s + 19$

   C. $19s$

   D. $9s + 10$

5. Jamal recorded the daily high temperatures over a two-week period.

   68°F, 75°F, 60°F, 69°F, 73°F, 74°F, 69°F, 70°F, 70°F, 65°F, 70°F, 71°F, 77°F, 71°F

   What is the interquartile range of the data?

   A. 4

   B. 5

   C. 6

   D. 17

6. Which is a coefficient in the expression $(3 \times 5) + 4x$?

   A. x

   B. 3

   C. 4

   D. 5

7. The table below shows how many of each flower Riley picked from her garden today.

Riley's Garden

Type of Flower	Number Picked
Tulips	12
Roses	6
Daffodils	8

Which ratio compares the number of daffodils picked to the number of tulips picked?

A. 1:2

B. 2:3

C. 3:4

D. 3:2

8. The surface area of a cube can be found with the formula $A = 6s^2$. If $s = 9$ cm, what is the surface area of the cube?

A. 54 cm^2

B. 81 cm^2

C. 486 cm^2

D. 729 cm^2

9. The expression $8a + 5c$ can be used to find the cost of an all-you-can-eat buffet lunch for *a* adults and *c* children. What is the cost of the lunch for 2 adults and 4 children?

A. $36

B. $42

C. $78

D. $136

10. Nora recorded her bowling scores for the month in the table below.

Bowling Scores

120	98	125	111	117
106	101	118	90	123
124	96	103	115	127

If she makes a histogram with intervals of 90–99, 100–109, 110–119, and 120–129, what is the height of the bar for the interval 120–129?

A. 1

B. 2

C. 3

D. 5

11. Which point on the number line represents −4?

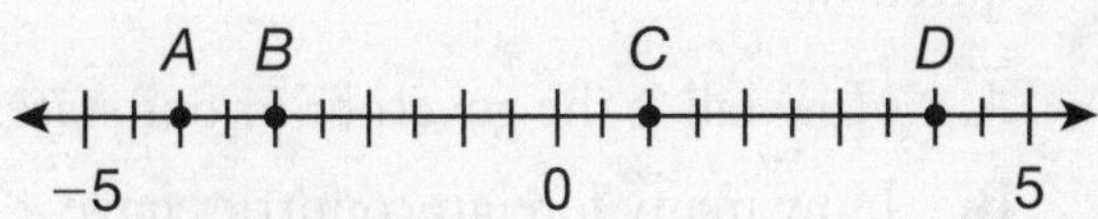

A. point *A*

B. point *B*

C. point *C*

D. point *D*

Summative Assessment

12. Bruce recorded the weights in pounds of the fish caught in a fishing contest. His data is shown below.

27, 34, 19, 8, 15, 32,
20, 24, 36, 30, 17

Make a box plot to display the data.

Weight of Fish (in pounds)

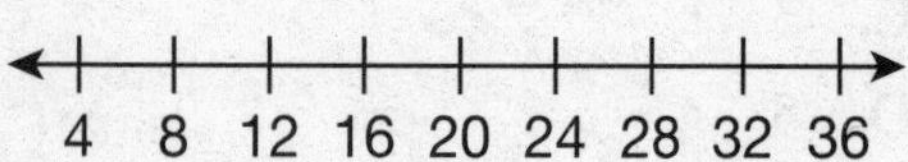

13. Which of the following is a statistical question?

A. How tall is the governor of our state?
B. How many foreign countries have you visited?
C. What is the temperature right now?
D. What are the colors of the U.S. flag?

14. Shawn is shopping for a rug for his dining room. The rectangular rug he likes measures 5 feet by 7 feet. What is the area of the rug?

A. 24 feet
B. 24 square feet
C. 35 feet
D. 35 square feet

15. Daniel rides his bicycle at a rate of 7 miles per hour. At that rate, how long would it take him to ride 28 miles?

A. 3 hours
B. 4 hours
C. 7 hours
D. 21 hours

16. Which is the opposite of -15?

A. 51
B. 15
C. -15
D. -51

17. Evan has 36 trading cards. This is 12 more trading cards than Max has. The equation $m + 12 = 36$ can be used to find how many trading cards Max has. How many trading cards does Max have?

A. 48
B. 24
C. 14
D. 3

18. Look at the figure below.

○○□□□□△△△△
○○□□□□△△△
○○□□□□△△△

For every 3 circles, how many triangles are there?

A. 2
B. 4
C. 5
D. 6

19. What is the product of 0.85×0.37?

A. 1.314
B. 0.673
C. 0.52
D. 0.3145

20. What expression represents seven less than five times a number?

A. $5 - 7a$
B. $7a - 5$
C. $7 - 5a$
D. $5a - 7$

21. Which point is located at $(-5, 2)$ on the coordinate grid?

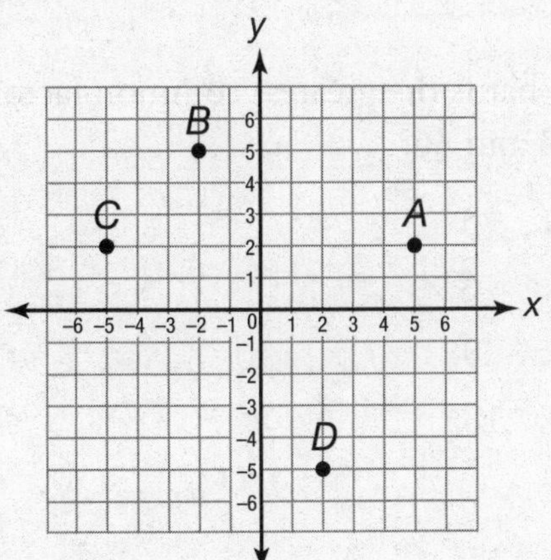

A. point A
B. point B
C. point C
D. point D

22. What is the area of this figure?

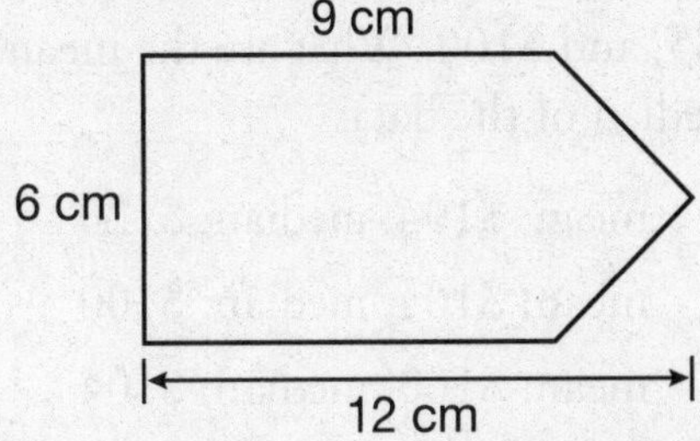

A. 42 cm^2
B. 63 cm^2
C. 66 cm^2
D. 72 cm^2

23. Which sentence is true?

A. $|-10| > |7|$
B. $|-8| < |-2|$
C. $|5| < |-1|$
D. $|-4| > |-6|$

24. Kendra has 9 trophies displayed on shelves in her room. This is $\frac{1}{3}$ as many trophies as Dawn has displayed. The equation $\frac{1}{3}d = 9$ can be use to find how many trophies Dawn has. How many trophies does Dawn have?

A. 3
B. 12
C. 27
D. 33

25. The daily profits for a bake sale at a school one week were \$120, \$95, \$120, \$85, and \$100. What are the mean and median of the data?

A. mean: \$104, median: \$120

B. mean: \$104, median: \$100

C. mean: \$100, median: \$104

D. mean: \$100; median: \$120

26. What solid figure can be made from this net?

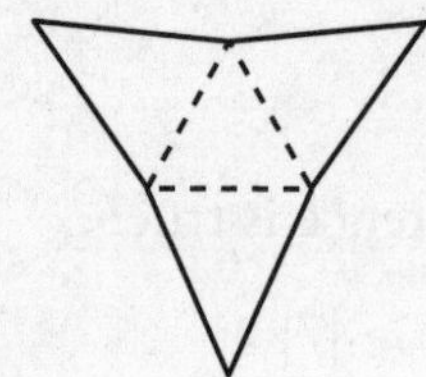

A. triangular pyramid

B. square pyramid

C. rectangular prism

D. triangular prism

27. A shirt company packs 24 shirts in a box. How many boxes do they need to pack 14,568 shirts?

A. 67

B. 607

C. 670

D. 6,007

28. The dot plot shows the number of days each student was absent from Miss Jordan's class during one month.

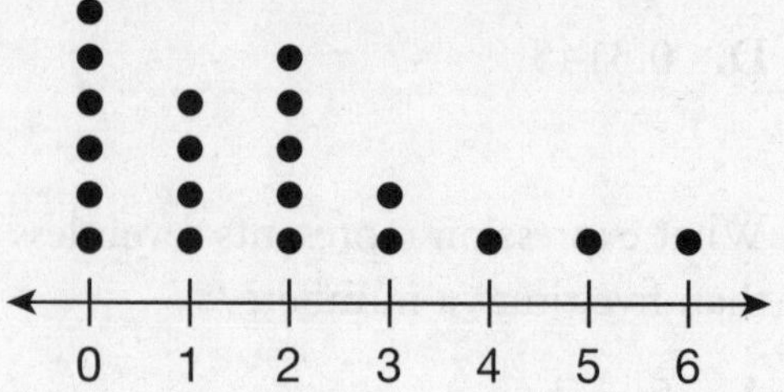

Which statement is **not** true?

A. The data clusters from 0 to 2 days.

B. The mode is 0 days.

C. There are 25 students in Ms. Jordan's class.

D. There is no gap in the data.

29. What is the greatest common factor of 63 and 36?

A. 3

B. 6

C. 9

D. 18

30. Shari made a net of a box to find how much wrapping paper she will need to wrap the box.

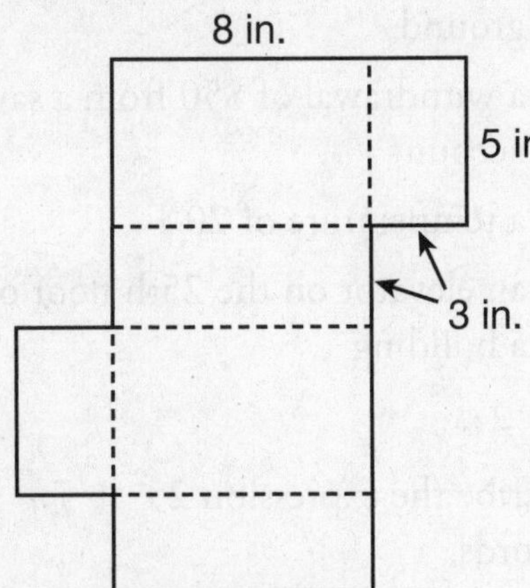

What is the least amount of wrapping paper she can use to cover all the surfaces?

A. 79 in.^{2}
B. 120 in.^{2}
C. 158 in.^{2}
D. 240 in.^{2}

31. Which shows how you can check that $\frac{7}{12} \div \frac{5}{8} = \frac{14}{15}$?

A. $\frac{14}{15} \div \frac{5}{8} = \frac{7}{12}$
B. $\frac{14}{15} \div \frac{7}{12} = \frac{5}{8}$
C. $\frac{5}{8} \times \frac{14}{15} = \frac{7}{12}$
D. $\frac{8}{5} \times \frac{14}{15} = \frac{7}{12}$

32. A recipe for snack mix has a ratio of 2 cups nuts, 4 cups pretzels, and 3 cups raisins. How many cups of nuts are there for each cup of raisins?

A. $\frac{2}{3}$ cup
B. $\frac{1}{2}$ cup
C. $\frac{1}{3}$ cup
D. $\frac{3}{4}$ cup

33. The table shows the number of chairs and tables needed for a banquet.

Tables and Chairs

Tables	2	3	4	?
Chairs	16	24	32	48

How many tables are needed for 48 chairs?

____

34. Which expression is equivalent to $7(n + 5)$?

A. $35n$
B. $n + 35$
C. $12n$
D. $7n + 35$

Summative Assessment

35. Twenty-eight sixth graders play basketball at J.R. Middle School. 20% of all sixth graders at J.R. play basketball. What is the total number of sixth graders at the school?

A. 140

B. 120

C. 90

D. 75

36. For which set of data would the median be best to describe the data?

A. the selling price of a loaf of bread

B. the weights of the boys on the wrestling team

C. the selling price of a used car

D. the number of students in each class at your school

37. The box plot shows the weights of newborn babies at a hospital.

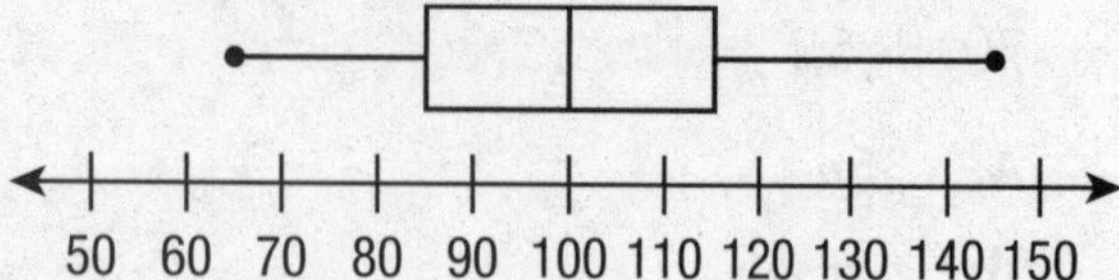

What is the median weight of the newborn babies, in pounds?

A. 4 pounds 1 ounce

B. 6 pounds 4 ounces

C. 7 pounds 3 ounces

D. 8 pounds 4 ounces

38. Which situation would you describe with a negative number?

A. a kite that is 30 feet above the ground

B. a withdrawal of $50 from a savings account

C. a temperature of 20°F

D. an elevator on the 25th floor of a building

39. Describe the expression $25^2 + 7n$ in words.

____

40. Which expression is equivalent to the expression below?

$$x + x + x$$

A. $x \times 2x$

B. $3 + x$

C. $3x$

D. $2(x + x)$

41. The mean absolute deviation (MAD) of the data below is 6.4.

24, 19, 16, 33, 33

How much greater than the MAD is the interquartile range of the data?

A. 9.1

B. 11.1

C. 17.6

D. 26.6

42. Plot and label point P at $(-5, 3)$ on the coordinate grid.

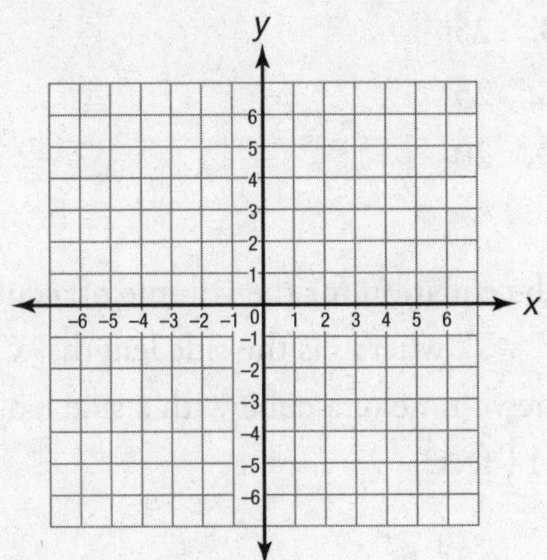

43. What is the quotient of $\frac{3}{4} \div \frac{5}{6}$?

A. $\frac{2}{3}$

B. $\frac{4}{5}$

C. $\frac{7}{8}$

D. $\frac{9}{10}$

44. Which expression represents the statement below?

6 times the sum of a number and 5

A. $6(5 + x)$

B. $x(5 + 6)$

C. $5x + 6$

D. $6x + 5$

45. One day, the low temperature in Minneapolis, Minnesota was 5°F below zero and the high temperature was 2°F above zero. Which of the following correctly compares these two temperatures?

A. $-5 > 2$

B. $-5 < -2$

C. $5 > 2$

D. $-5 < 2$

46. A recipe calls for 2 cups of water for every 5 cups of flour. How many cups of water are needed for 1 cup of flour?

A. $2\frac{1}{2}$ cups

B. 2 cups

C. $\frac{1}{2}$ cup

D. $\frac{2}{5}$ cup

47. Julia paid $120 for 6 gift cards. Each gift card was the same price. Which shows the equation that represents the situation and the price of each card?

A. $\frac{g}{6} = 120$; $720

B. $6g = 120$; $20

C. $g - 6 = 120$; $126

D. $g + 6 = 120$; $114

Summative Assessment

48. Which is equivalent to $|-30|$?

A. -30

B. -3

C. 3

D. 30

49. Raoul drew the trapezoid below.

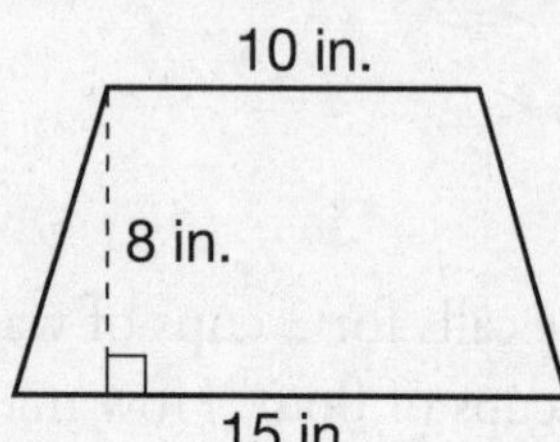

What is the area of the trapezoid?

____

50. The net of a square pyramid is shown below.

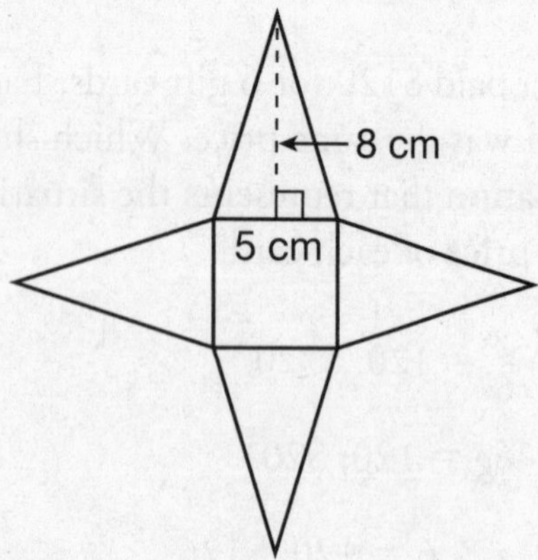

What is the surface area of the pyramid?

A. 80 cm^2

B. 100 cm^2

C. 105 cm^2

D. 185 cm^2

51. What is the quotient of 1,311 ÷ 57?

A. 26

B. 25

C. 23

D. 20

52. The equation for the volume of a cube is $V = s^3$, where s is the side length. What is the volume of a cube with a side length of $14\frac{1}{2}$ feet?

A. $43\frac{1}{2}$ ft^3

B. 2,744 ft^3

C. $3{,}048\frac{5}{8}$ ft^3

D. 3,375 ft^3

STOP

Session 2

53. Jeremy wants to make a map of his neighborhood using a coordinate grid. Each unit on the grid will equal 1 block.

A. Plot Jeremy's house at (5, 7), the library at (−2, 7), and the school at (5, −3).

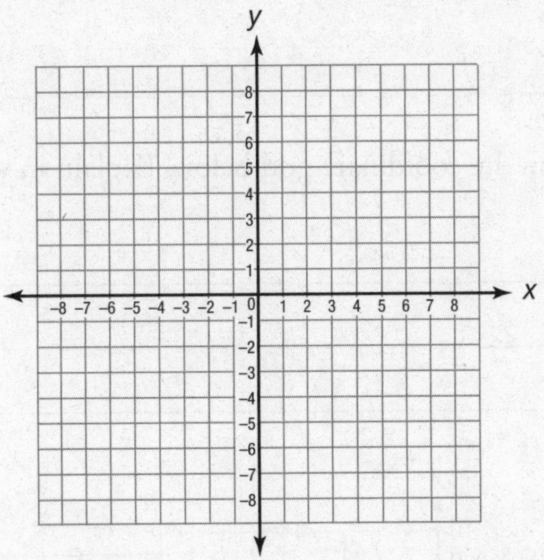

B. What is the distance, in blocks, from Jeremy's house to the school?

____

C. What is the distance, in blocks, from Jeremy's house to the library?

____

54. The equation $y = x - 4$ describes how the variables x and y are related.

A. Complete the table of values below for $y = x - 4$. Show all your work.

x	$y = x - 4$	y	(x, y)
4			
5			
6			
7			

B. Graph $y = x - 4$ on the coordinate grid below. Explain in words how you graphed the equation.

___

___

___

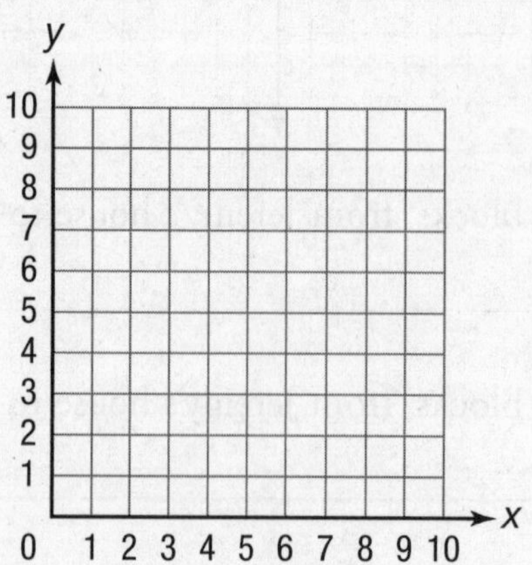

55. The number of laps around the track run by each student on the track team yesterday is shown below.

10, 15, 17, 16, 11, 18, 15, 16, 13, 16, 18

A. Make a dot plot of the data.

Number of Laps Ran

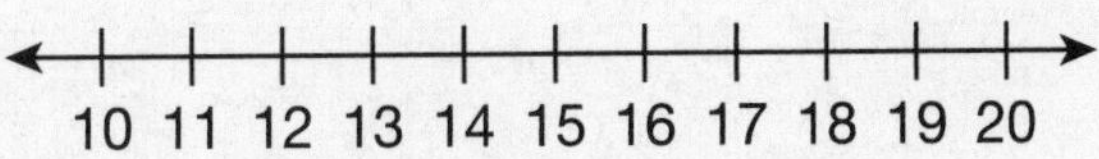

B. Find the interquartile range of the data set. Show your work.

C. How does the interquartile range reflect the data? Explain your answer.

___

___

___

Math Tools: Grid Paper

Math Tools: Coordinate Grid

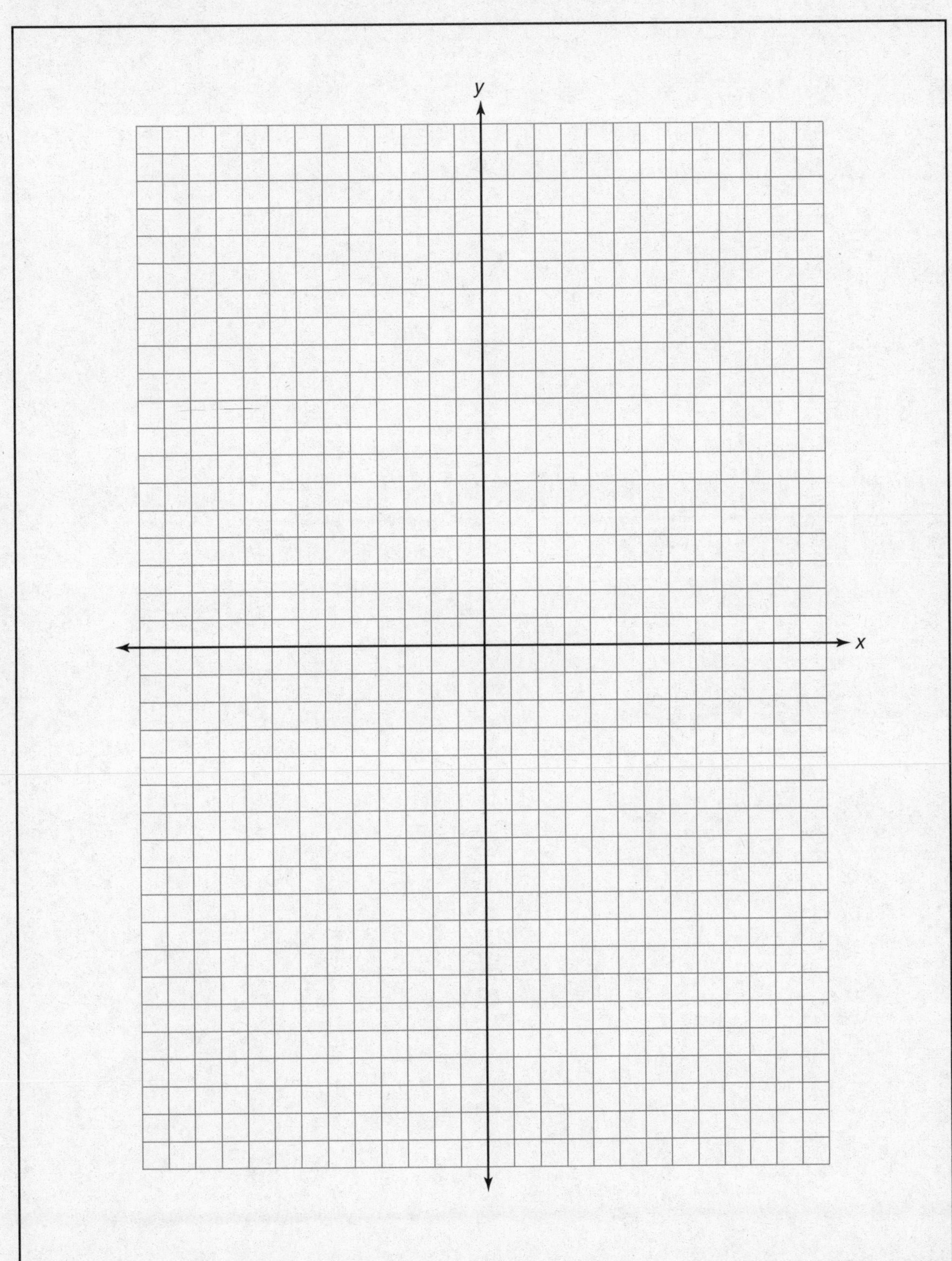

Cut or tear carefully along this line.

Math Tools: Coordinate Grid

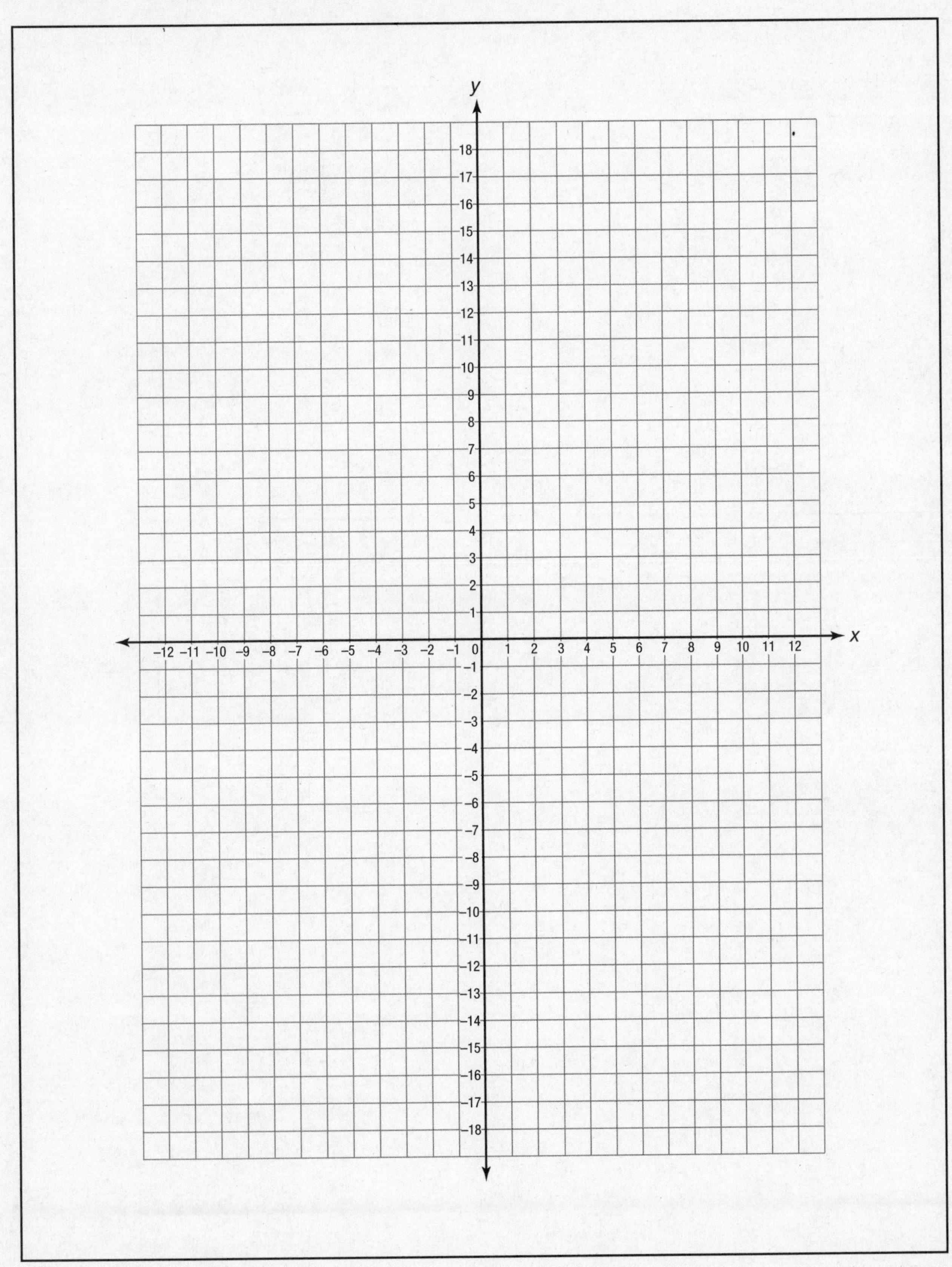

Math Tools: Grid Paper

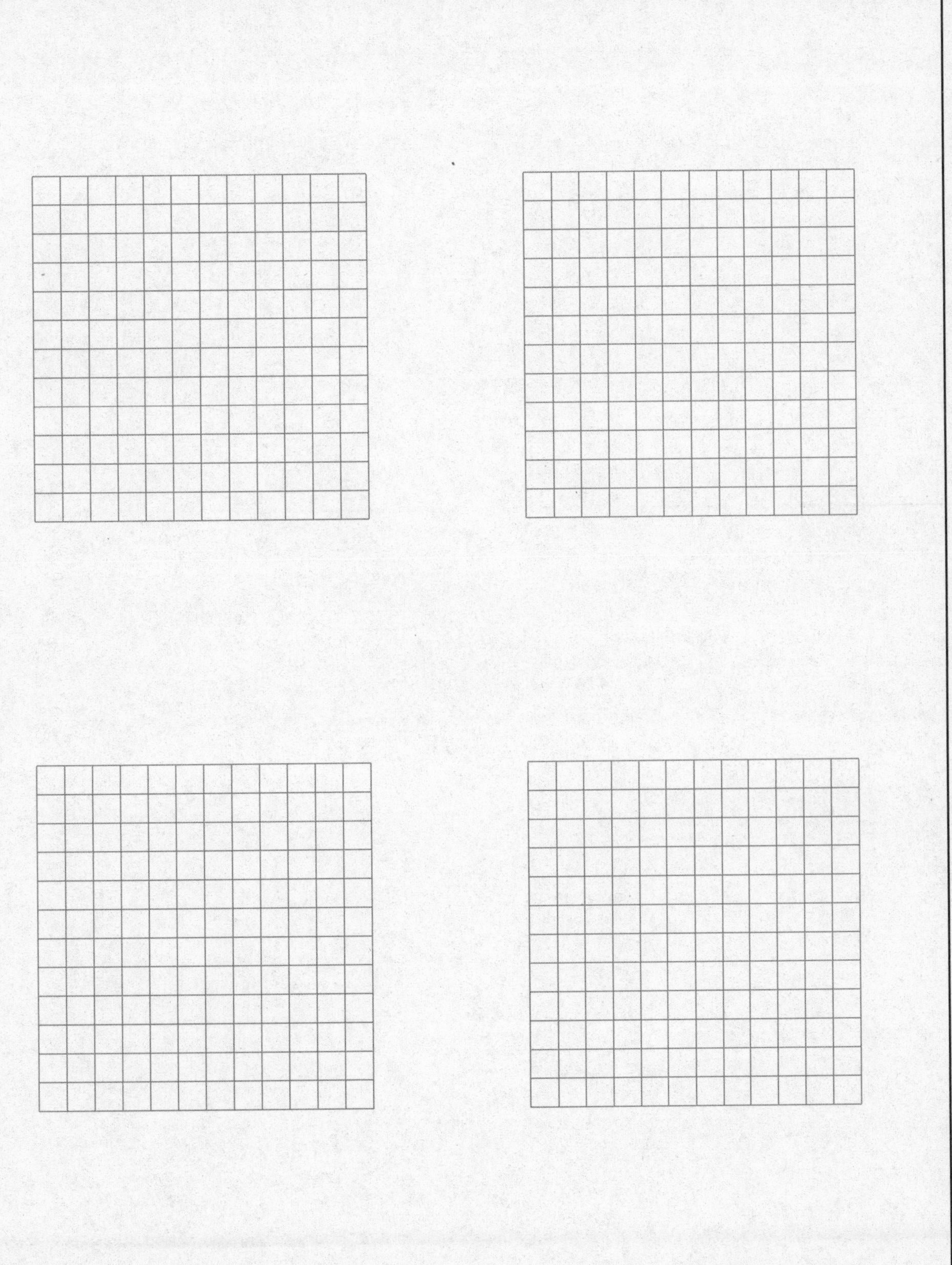

Math Tools: Coordinate Grid

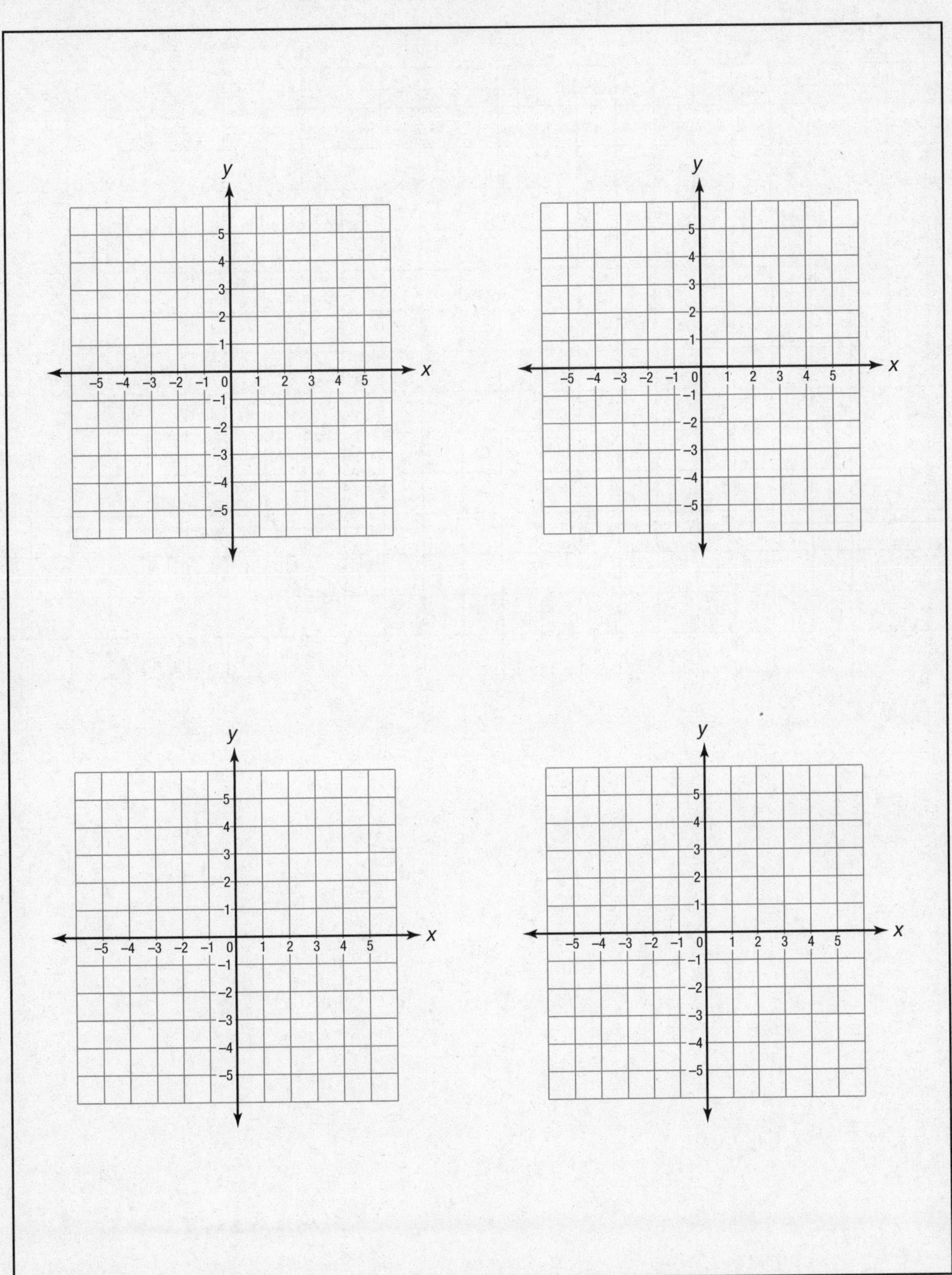

NOTES

NOTES

NOTES

NOTES

NOTES